Multi-terminal High-voltage Converter

Multi-terminal High-voltage Converter

Bo Zhang and Dongyuan Qiu
South China University of Technology
Guangzhou
China

Registered Offices
John Wiley & Sons, Inc., 111 River Street, Hoboken, NJ 07030, USA
John Wiley & Sons Singapore Pte. Ltd., 1 Fusionopolis Walk, #07-01 Solaris South Tower, Singapore 138628

Editorial Office
The Atrium, Southern Gate, Chichester, West Sussex, PO19 8SQ, UK

For details of our global editorial offices, customer services, and more information about Wiley products visit us at www.wiley.com.

Wiley also publishes its books in a variety of electronic formats and by print-on-demand. Some content that appears in standard print versions of this book may not be available in other formats.

Library of Congress Cataloging-in-Publication Data

Names: Zhang, Bo, 1962 October 23- author. | Qiu, Dongyuan, author.
Title: Multi-terminal high-voltage converter / Prof. Bo Zhang, Prof. Dongyuan
    Qiu.
Description: Hoboken, NJ : John Wiley & Sons, 2019. | Includes
    bibliographical references and index. |
Identifiers: LCCN 2018028413 (print) | LCCN 2018042477 (ebook) | ISBN
    9781119188360 (Adobe PDF) | ISBN 9781119188353 (ePub) | ISBN 9781119188339
    (hardcover)
Subjects: LCSH: Electric current converters. | Electric power
    distribution–High tension. | Electric power distribution–Direct current.
    | Electric power distribution–Alternating current.
Classification: LCC TK2796 (ebook) | LCC TK2796 .Z4233 2018 (print) | DDC
    621.31/3–dc23
LC record available at https://lccn.loc.gov/2018028413

Cover design by Wiley
Cover image: © iStock.com/kertlis

Set in 10/12pt WarnockPro by SPi Global, Chennai, India
Printed in Singapore by C.O.S. Printers Pte Ltd

10 9 8 7 6 5 4 3 2 1

Contents

About the Authors

Dr. Bo Zhang was born in Shanghai, China, in 1962. He received a BS degree in electrical engineering from Zhejiang University, Hangzhou, China in 1982, an MS degree in power electronics from Southwest Jiaotong University, Chengdu, China in 1988, and a PhD in power electronics from Nanjing University of Aeronautics and Astronautics, Nanjing, China in 1994.

He is currently a Professor at the School of Electric Power, South China University of Technology, Guangzhou, China. He has authored or coauthored three books, more than 450 papers, and 100 patents. His current research interests include nonlinear analysis and control of power electronics and AC drives.

Dr. Dongyuan Qiu was born in Guangzhou, China, in 1972. She received BSc and MSc degrees from the South China University of Technology, Guangzhou, China in 1994 and 1997, respectively, and a PhD from the City University of Hong Kong, Kowloon, Hong Kong in 2002.

She is currently a Professor at the School of Electric Power, South China University of Technology, Guangzhou, China. Her main research interests include the design and control of power converters, fault diagnosis, and sneak circuit analysis of power electronics.

Preface

The novel design and invention of power converter topologies are always a hot research topic in power electronics. Up to now, there have been thousands of power converter topologies proposed for different applications. It is found that the majority of these topologies, however, are derived and improved from the basic power electronic topologies. In terms of TRIZ (theory of the solution of inventive problems) proposed by the former Soviet inventor G. S. Altshuller, "the original inventions in any research field takes up only four percent of all, while the remaining ninety-six percent of inventions are the applications or combinations of them." Since power converter topologies design belongs to the area of invention, it follows the TRIZ principle without exception.

It is known that an MMC (modular multilevel converter) solves the problem of voltage sharing among power electronic devices under high voltage, while a nine-switch inverter realizes dual input or dual output of power electronic converters. Developed from MMC and the nine-switch converter, and combining the advantages of both, this book proposes the multi-terminal high-voltage converter. Besides, *Multi-terminal High-voltage Converter* has unique features, as follows:

1) It can simultaneously interface with several renewable sources such as hybrid systems of photovoltaic power generation and wind power generation.
2) It can simultaneously provide energy to multiple loads and the loads can be hybrid systems, including AC and DC loads.
3) It can be applied to AC–DC, DC–AC, DC–DC, and AC–AC units and some combinations as well as conversions of them.
4) It can form a novel high-voltage AC/DC transmission network and improve the performance of smart grids.

This book summarizes our achievements with power converter topologies in the past 10 years. We hope that it will provide new ideas for the topological invention of power electronic converters, and help researchers and engineers working on power electronics and smart grids to apply these topologies to AC/DC power transmission as a guide.

March 2018

Bo Zhang and Dongyuan Qiu
South China University of Technology
Guangzhou, China

Acknowledgments

Our research work was supported by the Team Program of the Natural Science Foundation of Guangdong Province (No. 2017B030312001), thus we must thank the Natural Science Foundation of Guangdong Province first for its funding. We would also like to thank our former graduate students, Mr. Jian Fu, Mr. Chong Han, Ms. Yuejuan Bian, and Mr. Li Qin, for their wonderful work in verifying our proposed topologies in multi-terminal high-voltage converters. We wish to express our sincere appreciation to the editors of this book, and to the staff of John Wiley & Sons Singapore for their professional and enthusiastic support of this project.

1

Overview of High-voltage Converters

1.1   Introduction

With the development of large-scale distributed generation (DG) systems access, such as photovoltaic (PV) panels, wind turbines, energy storage devices, and electric vehicles, the structure and features of power grids are bound to change, which will bring about not only a series of impacts on the safe and stable operation of power grids, but also challenges for ensuring power supply reliability and power quality.

In the traditional power grid shown in Figure 1.1a, the power flow is unidirectional in the power generation, transmission, substation, distribution, supply, and other sectors. Since the penetration rate of small-scale PV panels, electric vehicle charging stations, and energy storage stations is increasing in the power distribution side, as a result, the electric power is allowed to flow bidirectionally in the smart grid, as shown in Figure 1.1b.

As a smart grid is essentially different from a traditional power grid, the development of smart grids provides a great opportunity to interconnect different kinds of DG systems with the grid. In order to achieve a bidirectional flow of electricity and ensure the power quality of the grid, the power electronic converter must have good performance as an interface between the DG system and the power grid. Since a multilevel converter can extend the well-known advantages of low- and medium-power pulse-width modulation (PWM) converter technology into high-voltage high-power applications, it becomes the first choice for grid-connected interface converters and plays an important role in the following three aspects of smart grids:

1) *Power generation side.* The DC power generated by the large-scale PV station can be converted to AC power and transferred to the grid by a multilevel converter. The AC power from large wind farms can be converted into stable AC power and then connected to the grid by a two-stage multilevel converter. Large-scale storage plants can supply power to the grid and be charged by the grid via a bidirectional multilevel converter.
2) *Power transmission side.* High-voltage direct current (HVDC) power transmission associated with flexible alternating current transmission (FACT) are considered the most promising transmission technologies in smart grids. Since high-voltage high-power power conversion is often required, both HVDC and FACT systems were based initially on thyristor technology and more recently on fully controlled semiconductors and voltage-source multilevel converter topologies.

Multi-terminal High-voltage Converter, First Edition. Bo Zhang and Dongyuan Qiu.
© 2019 John Wiley & Sons Singapore Pte. Ltd. Published 2019 by John Wiley & Sons Singapore Pte. Ltd.

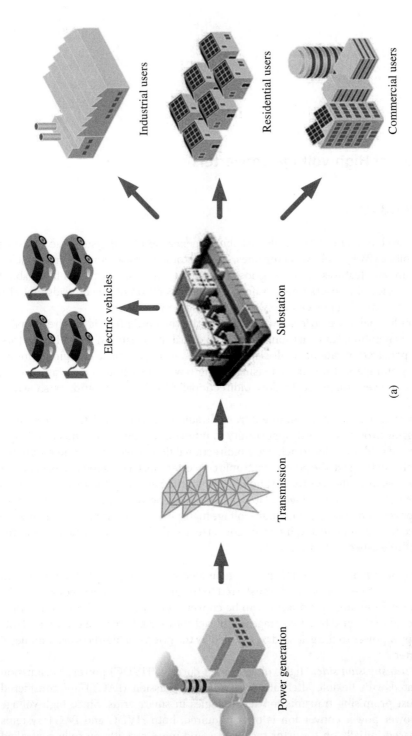

Power generation   Transmission   Substation

Electric vehicles

Industrial users

Residential users

Commercial users

(a)

Figure 1.1 Schematic of power grid. (a) Traditional power grid. (b) Smart grid.

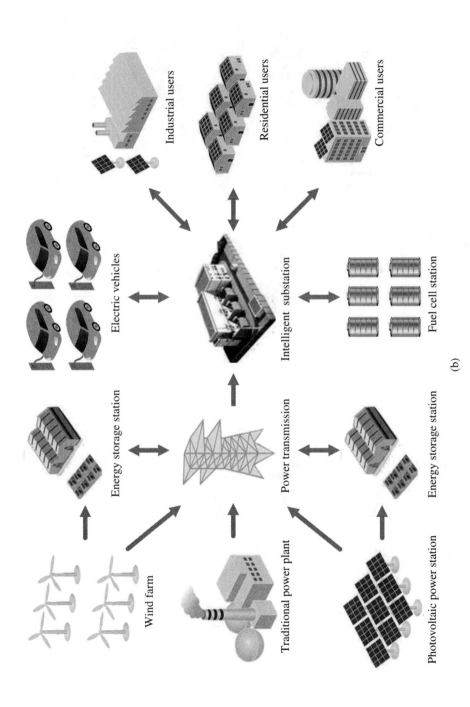

Figure 1.1 (*Continued*)

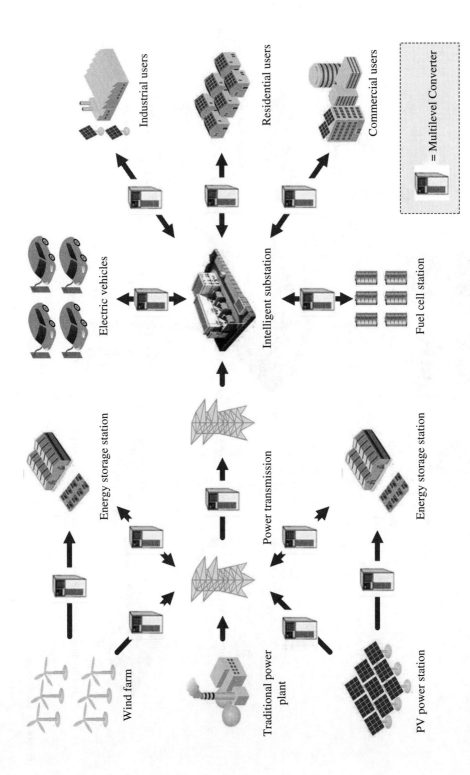

Figure 1.2 Application of multilevel converters in smart grid.

3) *Power distribution side.* The electric power generated by a small-scale wind and solar complementary system can be fed to the grid or supply directly to the user by multilevel converters. Making use of a bidirectional multilevel converter, the charging and discharging functions of the energy storage station or electric vehicle can be achieved.

In summary, a large number of multilevel converters can be found on the generation side, transmission side, and distribution side of a smart grid. As shown in Figure 1.2, future power grids will certainly be presented as an architecture with high-performance multilevel converters and high penetration of DG and energy storage systems.

On the power generation side, the large-scale DG systems, such as PV stations and wind farms, often consist of multiple subsystems or units. Although the multilevel converter effectively solves the high-voltage power conversion problem, the existing multilevel converter topologies are generally in the form of single input and single output; multiple multilevel converters are required to connect the large-scale DG system to the grid. Moreover, when more than one local load needs to be powered, only one multilevel converter can be used for one load, resulting in the need for multiple multilevel converters.

On the power distribution side of a smart grid, the power grid is not the only utility supplier to consumers. Other power suppliers include PV panels, wind turbines, fuel cells, supercapacitors, electric vehicles, and so on. Therefore, the interface converter needs to access multiple power sources and supply multiple loads. Obviously, the single-input single-output converter could not meet the above requirements; the use of multiple multilevel converters inevitably leads to a complex system configuration and high costs. Therefore, it is imperative to develop a high-voltage converter with multiple terminals.

In order to propose the architecture of a multi-terminal high-voltage converter, the development of a high-voltage high-power converter will be reviewed first, then several typical multilevel converters and common control schemes for multilevel converters will be introduced briefly.

1.2 Classification of High-voltage High-Power Converters

1.2.1 Two-Level Converters

High-voltage high-power converters have experienced high market penetration and noticeable development over the past two decades. The classical two-level converters were limited to low- or medium-power applications due to the blocking voltages of the power semiconductors with active turn-on and turn-off capabilities. The series connection of switching devices enabled the two-level converters to be applied in high-voltage high-power applications, while the number of switches in a series connection depends on the DC link voltage. As shown in Figure 1.3a, each phase of a typical high-power two-level voltage source inverter (2L-VSI) is composed of two groups of active switches, each consisting of three switches in series controlled by the same gating signal and hence working as a single switch [1]. In addition, additional capacitors in series could be necessary to reach the desired DC link voltage.

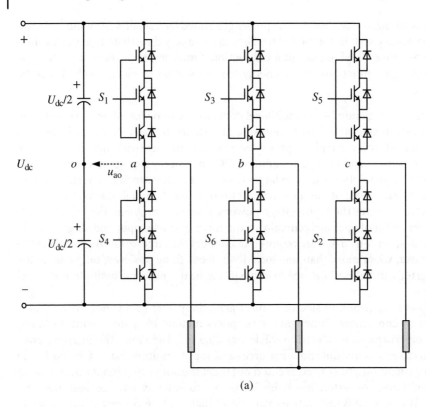

(a)

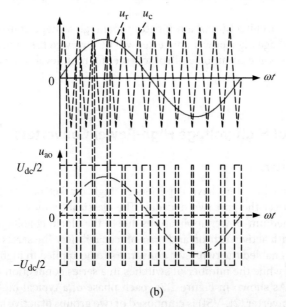

(b)

Figure 1.3 High-power two-level voltage source inverter. (a) Topology. (b) Bipolar PWM scheme.

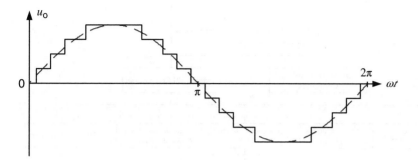

Figure 1.4 Typical stepped waveform of multilevel converter.

The most used modulation schemes for 2L-VSI are the well-known bipolar PWM, including third-harmonic injection, space-vector modulation (SVM), and selective harmonic elimination (SHE), which are usually used to enhance the quality of the output voltage. However, the challenge of the control strategy for high-power 2L-VSI is to require multiple switches per leg to operate at the same time. Moreover, only one DC level U_{dc} is utilized to create an average equal to the reference voltage in each switching cycle, as shown in Figure 1.3b; the switching loss and the total harmonic distortion (THD) of two-level converters are relatively high.

In order to synthesize the output voltage as close to a sinusoid as possible, a few additional components, like diodes or capacitors, can be added to the high-power 2L-VSI to generate a stepped waveform with less harmonic distortion, as shown in Figure 1.4, which originated the multilevel converter technology.

1.2.2 Multilevel Converters

Compared with the two-level waveform, the staircase waveform generated by the multilevel converter results in smaller dv/dt stress and lower THD, which can mitigate the problems associated with electromagnetic interference (EMI) and reduce the filter size. Moreover, the lower switching frequency and the lower voltage stress level for the switching devices lead to a significant reduction in switching losses. In general, the comparably lower switching losses and considerably higher power quality are the great advantages of multilevel converters compared with conventional two-level converters. Therefore, multilevel converters are the preferred choice for electric power conversion in high-voltage high-power applications using mature medium-voltage power semiconductor switches.

The classification of high-power converters is summarized in Figure 1.5. It is noted that only the voltage source converter topologies have been included, because the topic of this book is high-voltage converters. The most common voltage-source multilevel topologies are the neutral-point clamped (NPC) converter, flying capacitor (FC) converter, cascaded H-bridge (CHB) converter, and modular multilevel converter (MMC) [2]. Among these, the three-level neutral-point clamped converter (3L-NPC) presented by Nabae, Takahashi, and Akagi in 1980 is considered the first real multilevel power converter in medium-voltage applications [3]. Years later, the early concepts of the series-connected H-bridge (or CHB) and the FC circuit introduced in the 1960s developed into the multilevel converter topologies we know today [4, 5]. The topology

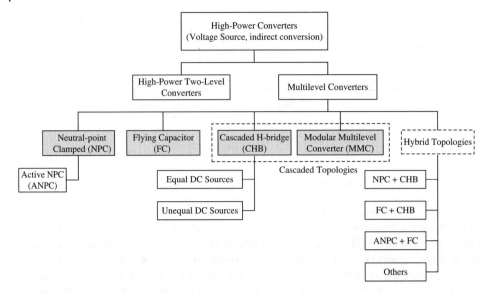

Figure 1.5 High-voltage converters classification.

of a modular multilevel converter (MMC or M2C) was developed in the early 2000s and has received increased attention since then [6].

It is obvious that these multilevel converters present different characteristics, such as the number of components, modularity, control complexity, efficiency, and fault tolerance. Depending on the application, the multilevel converter topology can be chosen by taking these factors into account. For completeness and better understanding of multilevel converter technology, several classic multilevel converter topologies will be reviewed in the next section.

1.3 Topologies of Multilevel Converters

1.3.1 Neutral-Point Clamped Converter

The first three-level NPC converter, also named the diode-clamped converter, was based on a modification of the classic two-level inverter by adding two additional power diodes per phase [3]. As shown in Figure 1.6a, the clamping diode is used to connect the neutral point N to the midpoint of two switches, then an additional voltage level "0" can be produced between the phase output point a and the neutral point N, when S_2 and $S_{1'}$ are switched ON. The phase output voltage u_{aN} will vary between 0 and $-U_{DC}/2$ or $U_{DC}/2$, which yields the name "three-level inverter."

A modified five-level neutral-diode clamped (5L-NPC) converter is shown in Figure 1.6b, in which both the main switches and the clamping diodes can be clamped [7]. In this five-level case, a total of 8 switches and 12 clamping diodes of equal voltage rating are used, and the DC bus consists of four storage capacitors. For a DC bus voltage U_{DC}, the voltage across each capacitor is $U_{DC}/4$, and each device's voltage stress is limited to one capacitor voltage level, or $U_{DC}/4$, through the clamping diodes.

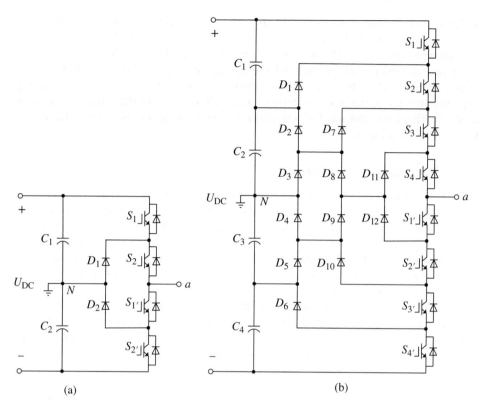

Figure 1.6 Phase structure of the NPC converter. (a) Three level. (b) Five level.

Table 1.1 Phase voltage of the 5L-NPC converter and its corresponding switch combinations.

Phase voltage u_{aN}	Switch states combination							
	S_1	S_2	S_3	S_4	$S_{1'}$	$S_{2'}$	$S_{3'}$	$S_{4'}$
$U_{DC}/2$	ON	ON	ON	ON	OFF	OFF	OFF	OFF
$U_{DC}/4$	OFF	ON	ON	ON	ON	OFF	OFF	OFF
0	OFF	OFF	ON	ON	ON	ON	OFF	OFF
$-U_{DC}/4$	OFF	OFF	OFF	ON	ON	ON	ON	OFF
$-U_{DC}/2$	OFF	OFF	OFF	OFF	ON	ON	ON	ON

The 5L-NPC converter is taken as an example to explain how the staircase voltage is synthesized; five different values for the phase voltage u_{aN} can be obtained using the switch combinations listed in Table 1.1. As a result, there will be nine levels for the phase-to-phase output: $\{-U_{DC}, -3U_{DC}/4, -U_{DC}/2, -U_{DC}/4, 0, U_{DC}/4, U_{DC}/2, 3U_{DC}/4, U_{DC}\}$.

By applying the above diode-clamped method to obtain a higher level, the synthesized output waveform adds more steps as the number of levels increases, producing a staircase wave which approaches the sinusoidal wave with minimum harmonic distortion.

Although the NPC structure can be extended to higher numbers of levels, it is less attractive because the number of devices will increase; for an m-level phase output, $m-1$ storage capacitors, $2(m-1)$ switches, and $(m-1)(m-2)$ clamping diodes are required. In particular, the clamping diodes, which have to be connected in series to block the higher voltages, will introduce more conduction losses and produce reverse recovery currents during commutation, affecting the switching losses of the other devices even more. Furthermore, the DC link capacitor voltage balance will become unfeasible at higher-level topologies.

1.3.2 Flying Capacitor Converter

The FC topology was first proposed by Meynard and Foch in 1992. As shown in Figure 1.7a, the switches in the three-level FC converter are arranged in two pairs (S_1, $S_{1'}$) and (S_2, $S_{2'}$), and the switches within each pair must always be in complementary states. The phase output voltage u_{aN} is equal to $U_{DC}/2$ when S_1 and S_2 are ON, $-U_{DC}/2$ when $S_{1'}$ and $S_{2'}$ are ON, 0 when S_1 and $S_{2'}$ or $S_{1'}$ and S_2 are ON. Obviously, there are redundant switch states in the FC converter that allow the switching stresses to be equally distributed between the switches [5].

For an m-level FC converter, its phase voltage has m levels while the line voltage has $2m-1$ levels, $(m-1)(m-2)/2$ FCs per phase, and $m-1$ capacitors for the DC bus are needed in total, and each capacitor as well as the switching device has the same voltage rating, that is $U_{DC}/(m-1)$. The phase structure of a five-level (5L) FC converter ($m=5$)

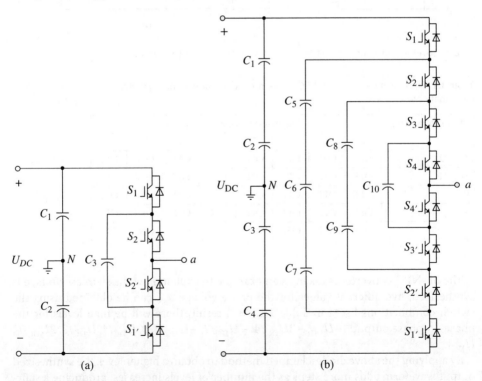

Figure 1.7 Phase structure of the FC converter. (a) Three level. (b) Five level.

Table 1.2 Phase voltage of the 5L-FC converter and its corresponding switch combinations.

Phase voltage u_{aN}	Switch states combination							
	S_1	S_2	S_3	S_4	$S_{1'}$	$S_{2'}$	$S_{3'}$	$S_{4'}$
$U_{DC}/2$	ON	ON	ON	ON	OFF	OFF	OFF	OFF
$U_{DC}/4$	ON	ON	ON	OFF	OFF	OFF	OFF	ON
	OFF	ON	ON	ON	ON	OFF	OFF	OFF
	ON	OFF	ON	ON	OFF	ON	OFF	OFF
	ON	ON	OFF	ON	OFF	OFF	ON	OFF
0	ON	ON	OFF	OFF	OFF	OFF	ON	ON
	OFF	OFF	ON	ON	ON	ON	OFF	OFF
	ON	OFF	ON	OFF	OFF	ON	OFF	ON
	ON	OFF	OFF	ON	OFF	ON	ON	OFF
	OFF	ON	OFF	ON	ON	OFF	ON	OFF
	OFF	ON	ON	OFF	ON	OFF	OFF	ON
$-U_{DC}/4$	ON	OFF	OFF	OFF	ON	ON	ON	ON
	OFF	OFF	OFF	ON	ON	ON	ON	OFF
	OFF	OFF	ON	OFF	ON	ON	OFF	ON
	OFF	ON	OFF	OFF	ON	OFF	ON	ON
$-U_{DC}/2$	OFF	OFF	OFF	OFF	ON	ON	ON	ON

is illustrated in Figure 1.7b, whose phase voltage can be synthesized by the switch combinations listed in Table 1.2 [8]. It is noticeable that the voltage synthesis in an FC converter has more flexibility than a diode-clamped converter.

Higher switching frequencies are necessary to keep the capacitors properly balanced, no matter what kind of capacitor balancing modulation is used, but these switching frequencies are not feasible for high-power applications, where they are usually limited to a range under 1 kHz. Besides the difficulty of balancing the voltage of capacitors connected in series, the major problem in the FC converter is the requirement for a large number of storage capacitors when the number of converter levels is high. Packaging the required number of bulky capacitors will become more difficult for high-level systems, as well as the initialization of the FC voltages.

1.3.3 Cascaded H-bridge Converter

The CHB converter consists of a series of H-bridge (or single-phase full-bridge) inverter units with separate DC sources, which are typically provided from a transformer/rectifier arrangement, or supplied from batteries, capacitors, or PV arrays. Figure 1.8 shows the single-phase configuration of a CHB converter with multiple H-bridge units, in which the AC terminal voltages of H-bridge units are connected in series. Thus, the phase output voltage is synthesized by the sum of inverter outputs, that is $u_{aN} = \sum_{i=1}^{N} u_{Hi}$.

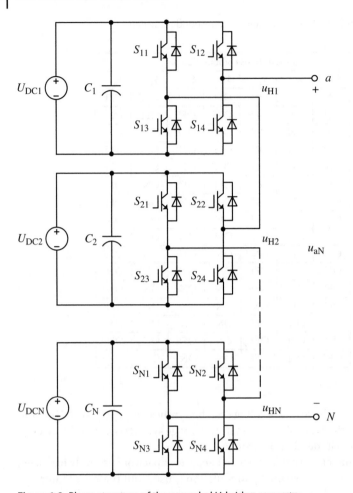

Figure 1.8 Phase structure of the cascaded H-bridge converter.

Assuming that the DC voltage source for the ith H-bridge unit is $U_{\mathrm{DC}i}$, the output voltage of the H-bridge, which is denoted by $u_{\mathrm{H}i}$, has three levels, $-U_{\mathrm{DC}i}$, 0, or $U_{\mathrm{DC}i}$, according to the switch combinations listed in Table 1.3. If the DC voltage of each unit is set to the same value ($U_{\mathrm{DC}i} = E$), then the level of the output phase voltage is defined by $m = 2N+1$ when N is the number of H-bridge units. If different DC voltages of the H-bridge unit are utilized, then a greater voltage level can be obtained for the phase output, which means that the power quality of the CHB converter may be greatly improved. For example, a CHB converter with two H-bridge units has five-level output with equal DC sources ($U_{\mathrm{DC}1} = U_{\mathrm{DC}2} = E$), but seven-level output with unequal DC sources ($U_{\mathrm{DC}2} = 2U_{\mathrm{DC}1} = 2E$), according to the voltage combinations in Table 1.4. Moreover, the 3L H-bridge unit in Figure 1.8 can be replaced by other kinds of bridge inverter, such as the 5L-NPC converter, which will maximize the number of voltage levels obtainable, resulting in high power quality [9].

Therefore, the primary advantage of the CHB converter is its modular structure that enables higher-voltage operation with classic low-voltage semiconductors; fewer or

Table 1.3 Output voltage of the H-bridge unit and its corresponding switch combinations.

Unit output voltage u_{Hi}	Switch states combination			
	S_{i1}	S_{i2}	S_{i3}	S_{i4}
U_{DCi}	OFF	ON	ON	OFF
0	ON	ON	OFF	OFF
	OFF	OFF	ON	ON
	OFF	OFF	OFF	OFF
$-U_{DCi}$	ON	OFF	OFF	ON

Table 1.4 Output voltage of the cascaded H-bridge converter when N = 2.

DC voltage source	Variables	Output voltage						
$U_{DC1} = U_{DC2} = E$	u_{H1}	E	E	0	0	−E	0	−E
	u_{H2}	E	0	E	0	0	−E	−E
	u_{aN}	2E	E	0	−E	−2E		
$U_{DC1} = 2E$	u_{H1}	2E	2E	0	0	0	−2E	−2E
$U_{DC2} = E$	u_{H2}	E	0	E	0	−E	0	−E
	u_{aN}	3E	2E	E	0	−E	−2E	−3E

more H-bridge units can be cascaded in order to decrease or increase the voltage and power level, respectively. However, the main disadvantage of this topology is that each H-bridge unit requires an isolated DC source. The whole system will be more expensive and bulky if the isolated DC sources are fed from phase-shifting isolation transformers.

1.3.4 Modular Multilevel Converter

The MMC, which is composed of multiple modules that are individually added up to synthesize the desired voltage, has become the most attractive multilevel converter topology for VSC-HVDC systems since its publication in 2003 [6].

The phase structure of MMC, consisting of 2N sub-modules (SMs), is shown in Figure 1.9, in which each arm comprises N series-connected SMs and a series inductor L. The upper (lower) arm of each phase-leg is represented by subscript "p" ("n"), the link between two inductors constituting the corresponding phase AC output. Normally, the cascaded SMs are identical and can be considered as a controlled voltage source whose maximum value is U_{SM}.

Since the high voltage at the DC side can be considered as two ideal DC voltage sources with amplitude $U_{DC}/2$, the limitation of the DC voltage and the phase output is related to the number of SMs per arm, that is $U_{DC}/2 + |u_{aN}| \leq N \cdot U_{SM}$. When $U_{SM} = U_{DC}/N$ is chosen, the phase output voltage u_{aN} is restricted to $-U_{DC} \leq u_{aN} \leq U_{DC}$ [10].

There are two kinds of common SM, one is the half-bridge sub-module (HBSM) shown in Figure 1.10a and the other is the full-bridge sub-module (FBSM) shown in Figure 1.10b [11]. The HBSM is composed of a DC capacitor and two switching devices, which are generally a unidirectional insulated-gate bipolar transistor (IGBT) and an

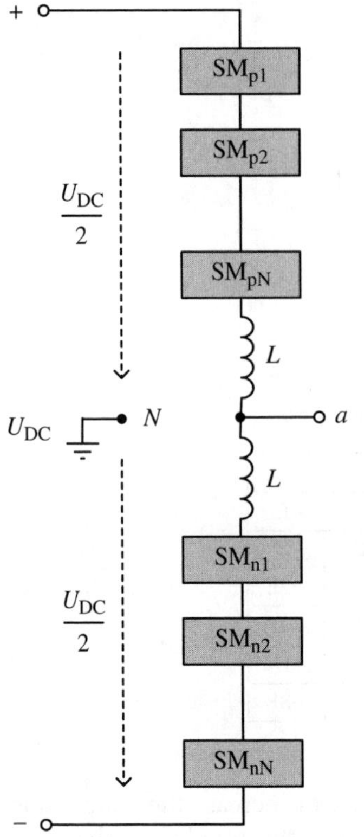

Figure 1.9 Phase structure of MMC.

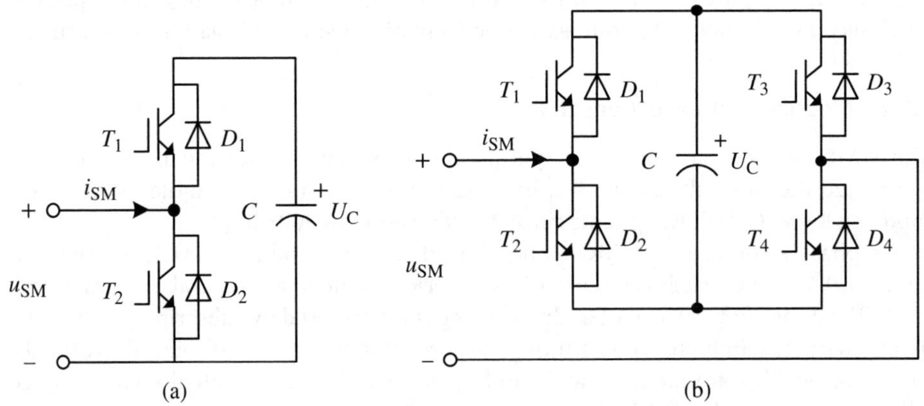

Figure 1.10 Circuit configurations of sub-module. (a) HBSM. (b) FBSM.

antiparallel diode. Two switching devices are driven by complementary signals, and the corresponding module output u_{SM} is either equal to its capacitor voltage U_C or zero.

The various states of the HBSM are listed in Table 1.5. These six states can be divided into three categories: capacitor ON, capacitor OFF, and energization states. The capacitor ON or inserted state appears when a high signal is applied to T_1, IGBT T_1, or the antiparallel diode D_1 conducts depending on the current direction of i_{SM}, and the entire

Table 1.5 Operating states of the HBSM.

No.	Switch ON	Switch OFF	State	Equivalent current path	u_{SM}	i_{SM}	Capacitor status
1							Positive	Charging
2	T_1	T_2	Capacitor ON/ Inserted		U_C		Negative	Discharging
3							Positive	Bypass
4	T_2	T_1	Capacitor OFF/ Bypassed		0		Negative	Bypass
5					U_C	Positive	Charging
6	—	T_1, T_2	Energization		0	Negative	Bypass

capacitor voltage U_C comes across the module terminals. The capacitor OFF or bypass state exists when the SM capacitor needs to be bypassed and have zero voltage across the terminals. To get this state, a high gate signal is applied to T_2. The energization state occurs when no signal is applied to either of the switches, and D_1 or D_2 conducts depending on the current direction of i_{SM}. This state never exists under normal operation of the converter, and only appears during converter failure. Obviously, by controlling the number of SMs in the capacitor ON (or OFF) state, the output voltage synthesized by the MMC is controlled. In addition, all capacitors in the MMC should be pre-charged to a nominal voltage of U_C [10].

The FBSM is composed of a DC capacitor and four switching devices; the output voltage u_{SM} depends on the switching states of T_1 to T_4. All operating states of the FBSM have been summarized in Table 1.6, states 1 to 4 belonging to "capacitor ON," states 5 to 8 belonging to "capacitor OFF," and states 9 and 10 being "energization."

Compared with the states of the HBSM in Table 1.5, there is one more capacitor ON state for the FBSM, in which the output voltage u_{SM} is equal to $-U_C$ when both T_2 and T_3 are switched ON. Another difference between the HBSM and the FBSM is the energization state; the capacitor C is inserted into the module terminals of the FBSM regardless of the current direction, which will be helpful in clearing DC faults. However, the power losses, as well as the cost of an MMC based on the FBSMs, are significantly higher than when using HBSMs, since the number of switching devices of an FBSM is twice that of an HBSM.

Based on the above analysis, the most distinctive advantages of MMC include: (i) its modularity and scalability to meet any voltage and power-level requirements; (ii) its low THD, which can reduce the size of passive filters; and (iii) its distributed location of capacitive energy storage, resulting in the absence of high-voltage DC link capacitors [11]. However, the disadvantage of the converter is the higher number of switching devices and gate units, and the total stored energy of the distributed capacitors is distinctly higher than that of other multilevel converters [12].

1.3.5 Active Neutral-Point Clamped Converter

The active neutral-point clamped (ANPC) converter is a derivative of the NPC converter presented in Section 1.3.1, which was proposed to overcome the extremely uneven distribution of conduction and switching losses [13]. As depicted in Figure 1.11, the three-level (3L) ANPC converter features additional active switches antiparallel to the clamp diode, which will bring new switch states and new commutations compared with the 3L-NPC shown in Figure 1.6a.

All switch states of the 3L-ANPC converter are given in Table 1.7. It can be found that the phase current can be conducted through the upper path of the neutral tap in both directions by turning on S_2 and S_3, while through the lower path of the neutral tap in both directions by turning on $S_{2'}$ and $S_{3'}$. As a result, there are four switch states for $u_{aN} = 0$, which could control the distribution of the switching losses during the commutations. Furthermore, $S_{3'}$ is turned on to guarantee an equal voltage sharing between $S_{1'}$ and $S_{2'}$ when $u_{aN} = U_{DC}/2$. Analogously, S_3 should be turned on when $u_{aN} = -U_{DC}/2$. Thus, the voltage-balancing resistors across the inner switches can be saved.

Table 1.6 Operating states of the FBSM.

No.	Switch ON	Switch OFF	Equivalent current path	u_{SM}	i_{SM}	Capacitor Status
1					Positive	Charging
	T_1, T_4	T_2, T_3		U_C		
2					Negative	Discharging
3					Positive	Discharging
	T_2, T_3	T_1, T_4		$-U_C$		
4					Negative	Charging
5					Positive	
	T_1, T_3	T_2, T_4				

(continued)

Table 1.6 (Continued)

No.	Switch ON	Switch OFF	Equivalent current path	u_{SM}	i_{SM}	Capacitor Status
6			T_1, D_1, T_3, D_3, T_2, D_2, T_4, D_4, C, i_{SM}, u_{SM}		Negative	
				0		Bypass
7	T_2, T_4	T_1, T_3	T_1, D_1, T_3, D_3, T_2, D_2, T_4, D_4, C, i_{SM}, u_{SM}		Positive	
8			T_1, D_1, T_3, D_3, T_2, D_2, T_4, D_4, C, i_{SM}, u_{SM}		Negative	
9	—	$T_1 \sim T_4$	T_1, D_1, T_3, D_3, T_2, D_2, T_4, D_4, C, i_{SM}, u_{SM}	U_C	Positive	Charging
10	—	$T_1 \sim T_4$	T_1, D_1, T_3, D_3, T_2, D_2, T_4, D_4, C, i_{SM}, u_{SM}	$-U_C$	Negative	Charging

Figure 1.11 Phase structure of the 3L-ANPC converter.

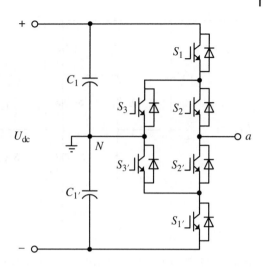

Table 1.7 Phase voltage of the 3L-ANPC converter and its corresponding switch combinations.

Phase voltage u_{aN}	Switch states combination					
	S_1	S_2	$S_{2'}$	$S_{1'}$	S_3	$S_{3'}$
$U_{DC}/2$	ON	ON	OFF	OFF	OFF	ON
0	OFF	ON	OFF	OFF	ON	OFF
	OFF	ON	OFF	ON	ON	OFF
	ON	OFF	ON	OFF	OFF	ON
	OFF	OFF	ON	OFF	OFF	ON
$-U_{DC}/2$	OFF	OFF	ON	ON	ON	OFF

It should be noted that ANPC with higher levels ($m > 3$) belongs to the hybrid multilevel converter, which will be introduced in the next section, because it cannot be obtained by the similar method used to construct the 3L-ANPC converter.

1.3.6 Hybrid Multilevel Converters

Since the introduction of the first multilevel topologies almost four decades ago, many multilevel converters have been published. However, most of them are variations on the three classic multilevel topologies, which are NPC, FC, and CHB, as discussed in the previous sections, or hybrids between them. Among the hybrid multilevel converter topologies, NPC+CHB [14], FC+CHB [15], and ANPC+FC [16] have received sustained attention because of their distinguished performances.

A. NPC+CHB

By cascading NPC and CHB phase legs, more different output voltage levels will inherently be produced. One kind of NPC+CHB topology is shown in Figure 1.12, which consists of a 3L-NPC with integrated-gate commutated thyristors (IGCTs) (main inverter) with an H-bridge (sub-inverter) with IGBTs in series [14], and the total level of phase output voltage will be $m = 9$. Normally, IGCTs with a high-voltage blocking capability

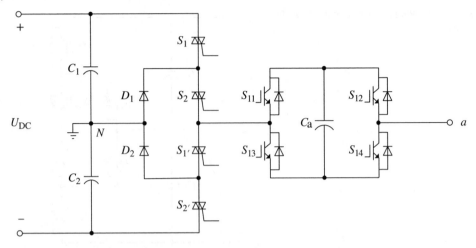

Figure 1.12 Phase structure of an NPC+CHB converter.

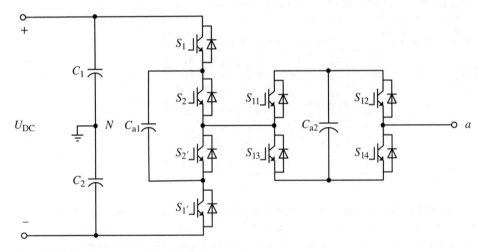

Figure 1.13 Phase structure of an FC+CHB converter.

are used to provide the main power with high reliability and low losses, while IGBTs with a higher switching-frequency capability are used to reduce the output harmonic content. To keep the system simple and the efficiency high, the sub-inverters are fed from the capacitors and supply reactive power only. Thus, the pre-charging of the capacitors in CHB becomes an issue to consider.

B. FC+CHB

FC+CHB is a cascaded topology consisting of a three-level FC converter and a capacitor-fed H-bridge in each phase, as shown in Figure 1.13 [15]. Compared with conventional topologies with the same voltage level, the key advantages of FC+CHB include a reduced number of components and the ability to balance the capacitor voltage by making use of the redundant switching states. Another important feature of this topology is its high reliability, because it can operate as a three-level inverter at full power rating even if one of the H-bridges fails and is bypassed.

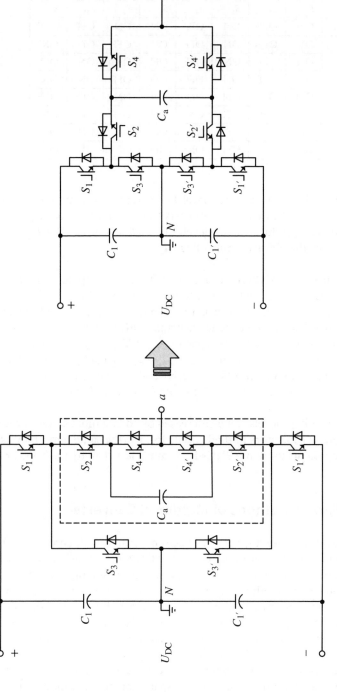

Figure 1.14 Phase structure of an ANPC+FC converter.

Table 1.8 Phase voltage of the 5L ANPC+FC converter and its corresponding switch combinations.

Phase voltage u_{aN}	Switch states combination							
	S_1	S_3	$S_{3'}$	$S_{1'}$	S_2	S_4	$S_{2'}$	$S_{4'}$
$U_{DC}/2$	ON	OFF	ON	OFF	ON	ON	OFF	OFF
$U_{DC}/4$	ON	OFF	ON	OFF	ON	OFF	OFF	ON
	ON	OFF	ON	OFF	OFF	ON	ON	OFF
0	ON	OFF	ON	OFF	OFF	OFF	ON	ON
	OFF	ON	OFF	ON	ON	ON	OFF	OFF
$-U_{DC}/4$	OFF	ON	OFF	ON	ON	OFF	OFF	ON
	OFF	ON	OFF	ON	OFF	ON	ON	OFF
$-U_{DC}/2$	OFF	ON	OFF	ON	OFF	OFF	ON	ON

If the DC bus voltage of the three-level FC converter in Figure 1.13 is U_{DC}, then the voltage across the H-bridge capacitor has to be maintained at $U_{DC}/4$. Obviously, this combination can produce voltage levels of $\{-3U_{DC}/4, -U_{DC}/2, -U_{DC}/4, 0, U_{DC}/4, U_{DC}/2, 3U_{DC}/4\}$ for the phase output voltage u_{aN}.

C. ANPC+FC

The common five-level ANPC converter is a hybrid of FC topology and NPC configuration [16], because its construction is different from that of 3L-ANPC in Section 1.3.5. As shown in Figure 1.14, the phase structure of a five-level ANPC+FC is derived from the 3L-ANPC by replacing the internal switching device pair (S_2, $S_{2'}$) by a 3L-FC cell. By choosing the voltage of the phase capacitor C_a as $U_{DC}/4$, five different voltage levels can be produced for the phase output: $\{-U_{DC}/2, -U_{DC}/4, 0, U_{DC}/4, U_{DC}/2\}$ [17]. All the allowed switching combinations of the 5L ANPC+FC are listed in Table 1.8, where the redundancies can effectively be utilized to control the neutral point and phase capacitor voltages.

The number of voltage levels can be increased by adding more series output switches and inserting phase capacitors in the FC cell [16]. For the same output levels, the total number of components used in APNC+FC is much less than that in NPC or FC.

1.4 Modulation Methods of Multilevel Converter

Multilevel converters present great advantages compared with conventional two-level converters; the improved quality and reduced THD of the output waveforms make multilevel converters very attractive to the industry. Correspondingly, multilevel converter modulation, and control methods, have attracted much attention, focusing on the challenge to extend traditional modulation methods to the multilevel case, the inherent complexity to control more power electronic devices, and the possibility of taking advantage of the extra switching states. As a consequence, a large number of different modulation and control methods have been developed.

As shown in Figure 1.15, the modulation methods for multilevel converters can be divided into two main groups, one based on space-vector generation and the other on

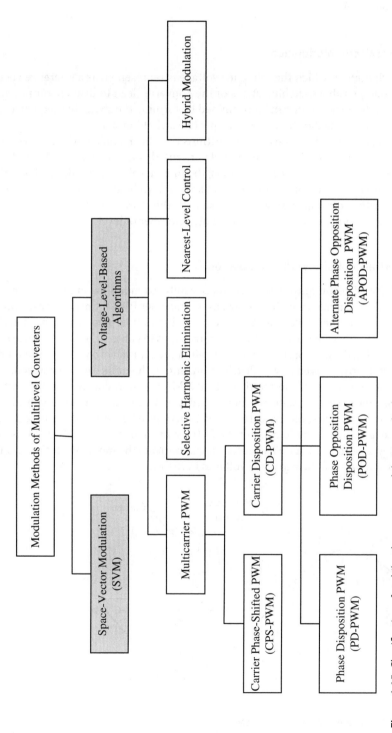

Figure 1.15 Classification of multilevel converter modulation methods.

voltage-level generation [18]. In this section, some typical modulation algorithms will be introduced briefly.

1.4.1 Space-Vector Modulation

SVM is a technique in which the reference voltage is represented as a reference vector. All the discrete possible switching states of the converter lead to discrete output voltages, but the reference vector can be combined by forming the switching sequence and calculating the on-state durations of the respective switching states.

In recent years, several space-vector algorithms have been applied to multilevel converters. Normally, most of them are designed particularly for a specific number of levels of the converter. In high-voltage applications, the more levels to be modulated by SVM, the more complex the vector calculation and selection to be done. The computational cost and algorithmic complexity will increase with the number of levels, thus SVM is not suitable for a multilevel converter with a high number of levels.

1.4.2 Multicarrier Pulse-Width Modulation

Traditional PWM techniques have been successfully extended for multilevel converters using multiple carriers, which is known as multicarrier PWM. Carrier phase-shifted pulse-width modulation (CPS-PWM) and carrier disposition pulse-width modulation (CD-PWM) are two widely used multicarrier strategies to control the MMC. Since each carrier is associated with a particular unit or cell in the multilevel converter and modulated independently, an even power distribution among the units can be provided.

For an m-level converter, the multicarrier PWM operation consists of $m-1$ different carriers, which have the same frequency and the same peak-to-peak amplitude. As shown in Figure 1.16, the symmetrical triangular carriers in CPS-PWM are phase shifted in a certain degree, for example, π/m for the CHB or $2\pi/m$ for the FC. CPS-PWM is normally employed with CHB and FC, because it can mitigate the input current harmonics in the CHB and balance the capacitor voltages in the FC.

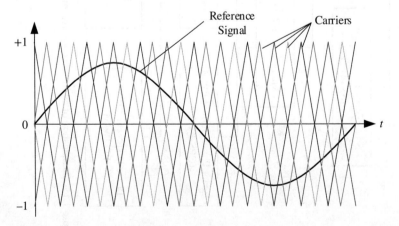

Figure 1.16 Modulation principle of CPS-PWM.

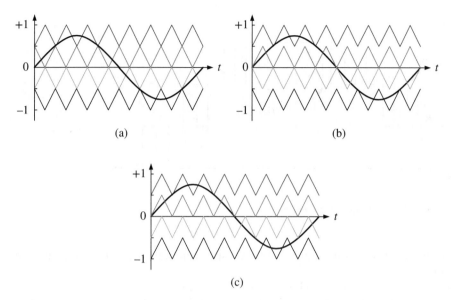

Figure 1.17 Modulation principle of different CD-PWMs. (a) PD-PWM. (b) POD-PWM. (c) APOD-PWM.

Different from CPS-PWM, the carriers in CD-PWM are arranged with shifts in amplitude. Depending on the disposition of the carriers, CD-PWM can be divided into phase disposition pulse-width modulation (PD-PWM), phase opposition disposition pulse-width modulation (POD-PWM), and alternate phase opposition disposition pulse-width modulation (APOD-PWM), as shown in Figure 1.17. The phase of each carrier is the same in PD-PWM, but all carriers are alternatively in phase opposition in APOD-PWM. In POD-PWM, all the carriers above the zero reference are in phase among them but in opposition with those below. The main difference among PD, POD, and APOD is in the harmonic content of the output PWM waveform. For POD and APOD, no harmonic exists at the carrier frequency due to the odd symmetry of their PWM waveforms. For the PD case, the waveform is asymmetric and the first set of undesired harmonics, or those aggregated near the carrier frequency, is relatively high. Thus, POD and APOD are more convenient for single-phase multilevel converters [19].

CD-PWM methods can be implemented for any multilevel topology, and are especially suitable for the NPC, since each carrier signal can easily be related to each power-switching device. However, in the case of low modulated ratio, the conduction time of the switching devices is inconsistent, which limits the further application of CD-PWM.

1.4.3 Selective Harmonic Elimination Modulation

In general, low-switching-frequency methods are preferred for high-power applications due to the reduction of switching losses, while high-switching-frequency algorithms are more suitable for better output quality and high-dynamic-range applications. Thus, selective harmonic elimination pulse-width modulation (SHE-PWM) is highly beneficial for high-power converters operating with low switching frequencies, owing

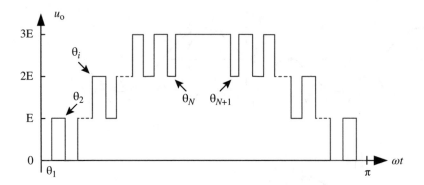

Figure 1.18 Multilevel waveform under SHE-PWM.

to its direct control of the output harmonic spectrum and the strong reduction in switching losses.

The concept of SHE-PWM techniques is based on the Fourier series decomposition of the periodic PWM waveform and calculation of the switching angles to eliminate selected low-order harmonics. Finding the analytical solution of the switching angles is the main challenge of SHE-PWM. Selection of a suitable solving algorithm or method relies heavily on the form of the PWM waveform, because waveform properties such as symmetry, the number and amplitude of voltage levels are important factors in determining the form and complexity of the solution [20]. A typical multilevel SHE-PWM waveform with quarter-wave symmetry is shown in Figure 1.18. Except for calculating the switching angles ($\theta_1 \sim \theta_N$), the distribution of these angles at different units or cells in a multilevel converter is an important aspect of the SHE-PWM method.

SHE-PWM becomes very complex to design and implement for converters with a high number of levels (above five), due to the increase in switching angles and hence nonlinear equations that need to be solved. In addition, online implementation relies on the need for very advanced computational tools and memory capabilities to accommodate the large lookup tables; SHE-PWM appears to be impractical for simultaneous compensation of multiple harmonics.

1.4.4 Nearest-Level Control Method

As presented in Section 1.4.1, multilevel SVM takes advantage of the voltage vectors generated by the converter to approximate the reference vector. The nearest-level control (NLC) method, in essence, has the same principle but considers the closest voltage level that can be generated by the converter instead of the closest vector. As shown in Figure 1.19, the NLC method generates a staircase output voltage by comparing the reference sine waveform with the settled voltage level.

Due to the simple concept and implementation of the NLC method, NLC is widely used in converters with a high number of levels. However, the THD of the output voltage will be high when the modulation index is low, because the operating principle of NLC is based on an approximation and not a modulation with a time average of the reference.

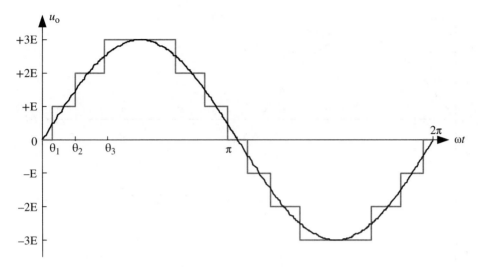

Figure 1.19 Multilevel waveform under NLC.

1.4.5 Hybrid Modulation

Hybrid modulation is in part a PWM-based method that is designed especially for a CHB with unequal DC voltage sources. Normally, the high-voltage units are controlled at the fundamental switching frequency, in which each switch is turned on and off only once per cycle, while the low-voltage unit is controlled by high-frequency PWM.

A CHB with two H-bridge units is taken as an example to illustrate the principle of hybrid modulation, and its typical outputs are shown in Figure 1.20.

As shown in Figure 1.20, the output of the unit with $U_{DC1} = 2E$ or u_{H1} is equal to 2E when the reference sine wave u_r is larger than E, −2E when $u_r < -E$, and otherwise $u_{H1} = 0$. The unipolar PWM scheme is applied to the unit with $U_{DC2} = E$, where the difference between u_{H1} and u_r is used as the reference. Obviously, the output voltage of the CHB is the seven-level waveform with low THD. Compared with the multicarrier PWM strategies, the switching losses of the high-voltage units are reduced by using hybrid modulation and the converter efficiency can be improved consequently.

1.5 Architecture of Multi-terminal High-voltage Converter

Nowadays, renewable energy systems (including wind and solar energies) are experiencing a sharp increase in use in the world; a multi-terminal direct current (MTDC) grid interconnecting multiple AC systems and offshore energy sources (e.g. wind farms) across nations and continents would allow effective sharing of intermittent renewable energy resources and open-market operation for secure and cost-effective supply of electricity. As a result, the need for multi-terminal high-voltage converters will grow rapidly.

In the majority of cases, multiple individual converters are required when multiple distributed AC power supplies need to be connected to the DC grid, which increases the

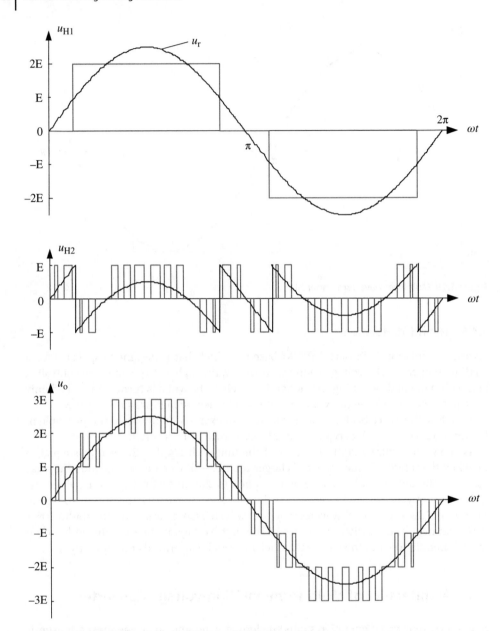

Figure 1.20 Outputs of a seven-level CHB under hybrid modulation.

overall complexity and hence cost of the system. In order to simplify the system structure with multiple AC terminals, it is necessary to design a novel independent high-voltage converter that can connect multiple AC sources simultaneously, as shown in Figure 1.21.

In practice, multi-terminal converters for independent control of large numbers of variable-speed electric drives are often required in industrial applications. Different topologies of multi-terminal converter with reduced number of semiconductor devices

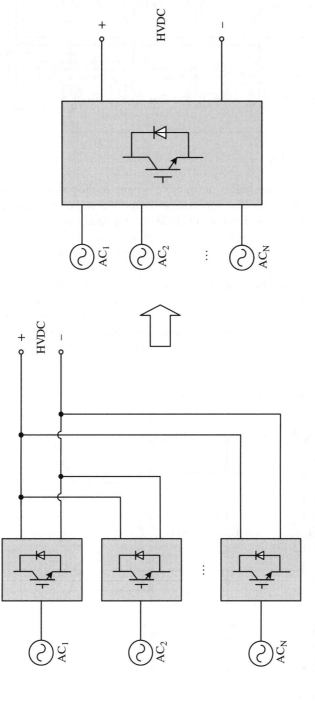

Figure 1.21 Architecture of a multi-terminal high-voltage converter.

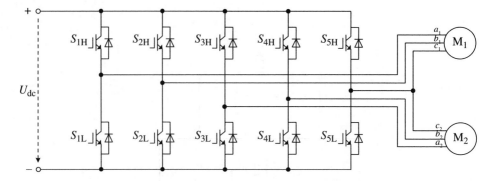

Figure 1.22 Five-leg inverter.

have been developed for the multi-motor drive systems in the past two decades, in order to realize further reductions in complexity and capital cost. For example, the standard dual three-phase voltage source inverter (VSI) configuration is usually used to supply two three-phase induction machines; typical single inverters that can drive two AC motors independently include the five-leg inverter [21], nine-switch inverter [22], and dual two-phase inverter [23].

The five-leg inverter, which has 10 switches, is shown in Figure 1.22; one inverter leg is common to both machines, while the other four inverter legs are connected to the phases of one machine only. The structure of the nine-switch inverter is shown in Figure 1.23, which combines two three-phase inverters with three common switches. The upper inverter for motor M_1 consists of switches S_{UH}, S_{VH}, S_{WH}, S_{UM}, S_{VM}, and

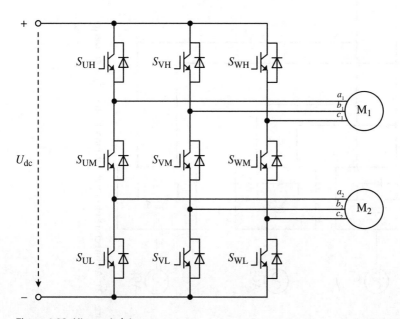

Figure 1.23 Nine-switch inverter.

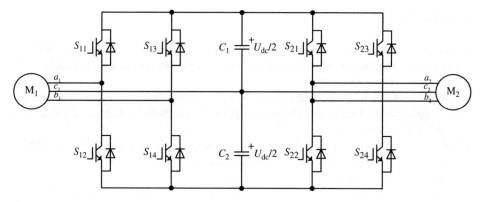

Figure 1.24 Dual two-phase inverter.

S_{WM}. The lower inverter for motor M_2 consists of switches S_{UM}, S_{VM}, S_{WM}, S_{UL}, S_{VL}, and S_{WL}. As shown in Figure 1.24, two two-phase inverters are connected back-to-back with a common DC bus, thus requiring only 8 switches instead of the conventional 12. Similar to the five-leg inverter, two of the machine's phases are connected to two legs of the inverter, but the third phase is connected to the center point of the DC link capacitors.

Therefore, based on the idea of minimizing components in the above converters, it is possible to construct an independent multi-terminal high-voltage converter by extending the terminal number from two to multiple. Taking the needs of high-voltage high-power into account, a multi-terminal DC–AC high-voltage inverter topology is put forward for the first time in this book by combining the nine-switch inverter with the MMC converter. A similar method has been applied to construct a multi-terminal AC–DC high-voltage converter, a multi-terminal AC–AC high-voltage converter, a multi-terminal DC–DC high-voltage converter, and so on [24].

1.6 Arrangement of this Book

The target of this book is to offer an overview of the existing technology and future trends in high-voltage converters, with discussion and analysis of multi-terminal high-voltage converters. Thus, the book starts with an introductory chapter about various kinds of existing high-voltage converter. As the development trend of modern power electronic systems is to pursue low cost, fewer components, high efficiency, and high reliability, a new kind of high-voltage converter with bridge module is proposed in Chapter 2, which is followed by a series of novel multi-terminal high-voltage converters in Chapters 3–7. The proposed multi-terminal high-voltage converters can realize different kinds of conversion, including single DC input/multiple AC outputs, multiple AC inputs/single DC output, multiple AC inputs/multiple AC outputs, multiple DC inputs/multiple DC outputs, and hybrid conversion. The following two chapters, Chapters 8 and 9, focus on some common issues and potential industrial applications of the proposed multi-terminal high-voltage converters, such as short-circuit protection and capacitor voltage balancing schemes.

References

1 Rodríguez, J., Bernet, S., Wu, B. et al. (2007). Multilevel voltage-source-converter topologies for industrial medium-voltage drives. *IEEE Transactions on Industrial Electronics* 54 (6): 2930–2945.

2 Kouro, S., Malinowski, M., Gopakumar, K. et al. (2010). Recent advances and industrial applications of multilevel converters. *IEEE Transactions on Industrial Electronics* 57 (8): 2553–2580.

3 Nabae, A., Takahashi, I., and Akagi, H. (1981). A new neutral-point-clamped PWM inverter. *IEEE Transactions on Industry Applications* IA-17 (5): 518–523.

4 Marchesoni, M., Mazzucchelli, M., and Tenconi, S. (1990). A nonconventional power converter for plasma stabilization. *IEEE Transactions on Power Electronics* 5 (2): 212–219.

5 Meynard, T. A., Foch, H. (1992) Multilevel conversion: high voltage choppers and voltage-source inverters. Record of 23rd Annual IEEE Power Electronics Specialists Conference, pp. 397–403.

6 Lesnicar, A., Marquardt, R. (2003) An innovative modular multilevel converter topology suitable for a wide power range. IEEE Bologna Power Tech Conference Proceedings, Vol. 3, pp. 1–6.

7 Yuan, X. and Barbi, I. (2000). Fundamentals of a new diode clamping multilevel inverter. *IEEE Transactions on Power Electronics* 15 (4): 711–718.

8 Lai, J. and Peng, F. (1996). Multilevel converters – a new breed of power converters. *IEEE Transactions on Industry Applications* 32 (3): 509–517.

9 Corzine, K. and Familiant, Y. (2002). A new cascaded multilevel H-bridge drive. *IEEE Transactions on Power Electronics* 17 (1): 125–131.

10 Sahoo, A., Otero-De-Leon, R., Mohan N. (2013) Review of modular multilevel converters for teaching a graduate-level course of power electronics in power systems. Proceedings of 2013 North American Power Symposium (NAPS), pp. 1–6.

11 Debnath, S., Qin, J., Bahrani, B. et al. (2015). Operation, control, and applications of the modular multilevel converter: a review. *IEEE Transactions on Power Electronics* 30 (1): 37–53.

12 Rohner, S., Bernet, S., Hiller, M. et al. (2010). Modulation, losses, and semiconductor requirements of modular multilevel converters. *IEEE Transactions on Industrial Electronics* 57 (8): 2633–2642.

13 Brückner, T. and Bernet, T. (2005). The active NPC converter and its loss-balancing control. *IEEE Transactions on Industrial Electronics* 52 (3): 855–868.

14 Veenstra, M. and Rufer, A. (2005). Control of a hybrid asymmetric multilevel inverter for competitive medium-voltage industrial drives. *IEEE Transactions on Industry Applications* 41 (2): 655–664.

15 Roshankumar, P., Rajeevan, P., Mathew, K. et al. (2008). A five-level inverter topology with single-DC supply by cascading a flying capacitor inverter and an H-bridge. *IEEE Transactions on Power Electronics* 27 (8): 3505–3512.

16 Barbosa, P., Steimer, P., Steinke, J. et al. (2005). Active neutral-point-clamped multilevel converters. *Proceedings of IEEE Power Electronic Specialist Conference* 2296–2301.

17 Oikonomou, N., Karamanakos, P., and Kieferndorf, F. (2013). Model predictive pulse pattern control for the five-level active neutral-point-clamped inverter. *IEEE Transactions on Industry Applications* 49 (6): 2058–2592.

18 Franquelo, L. G., Rodriguez, J. L. J. et al. (2008). The age of multilevel converters arrives. *IEEE Industrial Electronics Magazine* 2 (2): 28–39.

19 Naderi, R. and Rahmati, A. (2008). Phase-shifted carrier PWM technique for general cascaded inverters. *IEEE Transactions on Power Electronics* 23 (3): 1257–1269.

20 Dahidah, M., Konstantinou, G., and Agelidis, V. (2015). A review of multilevel selective harmonic elimination PWM: formulations, solving algorithms, implementation and applications. *IEEE Transactions on Power Electronics* 30 (8): 4091–4106.

21 Francois, B., Bouscayrol, A. (1999) Design and modeling of a five-phase voltage-source inverter for two induction motors. Proceedings of Europe Conference Power Electronics and Application (EPE), p. 626.

22 Ledezma, E., McGrath, B., Munoz, A., and Lipo, T. A. (2001). Dual AC-drive system with a reduced switch count. *IEEE Transactions on Industry Application* 37 (5): 1325–1333.

23 Kominami, T. and Fujimoto, Y. (2007). A novel nine-switch inverter for independent control of two three-phase loads. *IEEE Industrial Applications Conference* 2346–2350.

24 Zhang, B., Qiu, D. Y., and Fu, J. (2015). Topology and analysis of novel multi-terminal high-voltage converters. *Journal of Power Supply* 13 (6): 69–76.

2

Multiple-Bridge-Module High-voltage Converters

2.1 Introduction

The modular multilevel converter (MMC) introduced in Section 1.3.4 has gained in popularity, especially in high-voltage direct current (HVDC) and flexible alternating current transmission systems (FACTS). Normally, an MMC system may include hundreds or even thousands of sub-modules; if more transmitted power or a higher voltage level is needed, the number of sub-modules will increase accordingly. As shown in Figure 1.10, every sub-module contains one capacitor. It is reported that the capacitor usually accounts for over 50% of the total volume and 80% of the weight, while its cost is almost equal to the total cost of the power switches. In order to reduce the construction and maintenance costs, it is desirable to reduce the number of capacitors in the system.

As the capacitor voltage of every sub-module in the MMC is controlled to have the same value, it is supposed that sub-modules in different phases could share one common capacitor. Therefore, a new type of bridge module is proposed by combining two or three bridge cells and one capacitor together. Similar to the structure of the MMC, a novel high-voltage converter named the multiple-bridge-module converter (MBMC) is put forward for the first time by connecting the bridge modules in series. In this chapter, the topological structure, operating principle, and control scheme of the different kinds of MBMC will be discussed in detail.

2.2 Configuration of Bridge Module

It is known that each arm of the MMC is made up of N sub-modules connected in series. If the ith sub-modules in different legs are integrated into one module, then a new kind of module is obtained and named the "bridge module." According to the common type of sub-module in the MMC, only the half-bridge cell and the full-bridge cell are used to construct the bridge module in this section.

Multi-terminal High-voltage Converter, First Edition. Bo Zhang and Dongyuan Qiu.

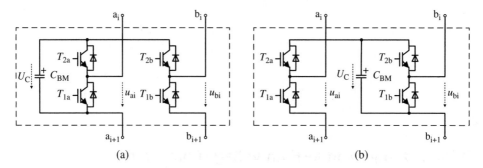

Figure 2.1 Single-phase half-bridge module. (a) Capacitor in leg *a*. (b) Capacitor in leg *b*.

2.2.1 Half-Bridge Module

Figure 2.1 illustrates the single-phase half-bridge module (SP-HBM), which includes two half-bridge cells and one common capacitor C_{BM}. The switching device in the half-bridge cell could provide bidirectional current flow, since there is an antiparallel diode in parallel with the power switch. Capacitor C_{BM} can be connected in parallel with the half-bridge cell in either leg *a* or leg *b*, as shown in Figure 2.1a,b, respectively.

The single-phase half-bridge module in Figure 2.1a is taken as an example. When the upper switch T_{2a} of leg *a* is turned ON and the lower switch T_{1a} is turned OFF, the voltage between terminals a_i and a_{i+1} is determined by the capacitor voltage U_C, that is $u_{ai} = U_C$, and the corresponding status is named "1." When T_{2a} is turned OFF and T_{1a} is turned ON, terminals a_i and a_{i+1} are shorted, and the capacitor C_{BM} is bypassed, which results in $u_{ai} = 0$, then the corresponding status is named "0." For leg *b*, $u_{bi} = 0$ or the status "0" happens when T_{1b} is turned ON and T_{2b} is turned OFF. However, when T_{1b} is turned OFF and T_{2b} is turned ON, terminal b_i is connected to the positive terminal of capacitor C_{BM} while terminal b_{i+1} is separated from the other components in the same bridge module. At this time, the value of u_{bi} is uncertain, because it is not only determined by the status of leg *a* in the same bridge module, but also by the status of leg *b* in other bridge modules. In order to avoid the complex derivation of u_{bi}, T_{2b} can be turned OFF as well as T_{1b}. Thus, the above two statuses, when T_{1b} is OFF, are represented by "X_1" and "X_2," respectively.

The switching states and terminal voltages of SP-HBM under different status are summarized in Table 2.1. It is obvious that only the terminal voltage u_{ai} can be controlled by the ON/OFF state of the power switches when C_{BM} is in parallel with leg *a*.

Table 2.1 Switching states and terminal voltages of SP-HBM.

Leg *m*	Status	T_{1m}	T_{2m}	u_{mi}
a (leg with parallel capacitor)	1	OFF	ON	U_C
	0	ON	OFF	0
b (leg without parallel capacitor)	X_1	OFF	ON	—
	X_2	OFF	OFF	
	0	ON	OFF	0

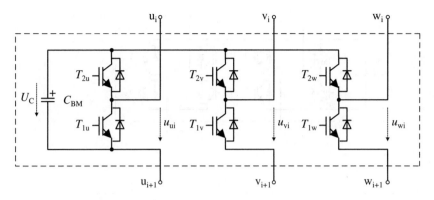

Figure 2.2 Three-phase half-bridge module.

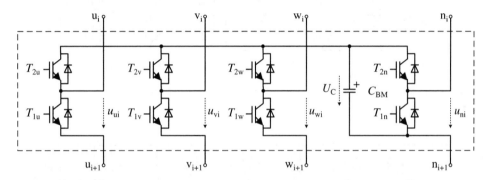

Figure 2.3 Three-phase four-leg half-bridge module.

Considering the transformer winding connection, there are two types of three-phase power distribution system, one is three-phase three-wire and the other is three-phase four-wire. Based on the structure of the single-phase bridge module in Figure 2.1, the three-phase half-bridge module (TP-HBM) for the three-phase three-wire system is illustrated in Figure 2.2, in which the common capacitor is on leg u. It should be noted that the capacitor can be connected to the half-bridge cell in leg v or leg w as well.

For three-phase four-wire applications, the three-phase four-leg half-bridge module (TPFL-HBM) can be obtained by adding one more leg to the TP-HBM. It is known that the position of the capacitor will determine which leg voltage is controllable; the terminal voltage of the leg with parallel capacitor will be different from those without a capacitor. In order to obtain symmetry of leg voltages or phase voltages, the capacitor in the TPFL-HBM must be connected to the neutral leg, which is shown in Figure 2.3.

2.2.2 Full-Bridge Module

By replacing the half-bridge cell in the bridge module with the full-bridge one, the single-phase full-bridge module (SP-FBM), the three-phase full-bridge module

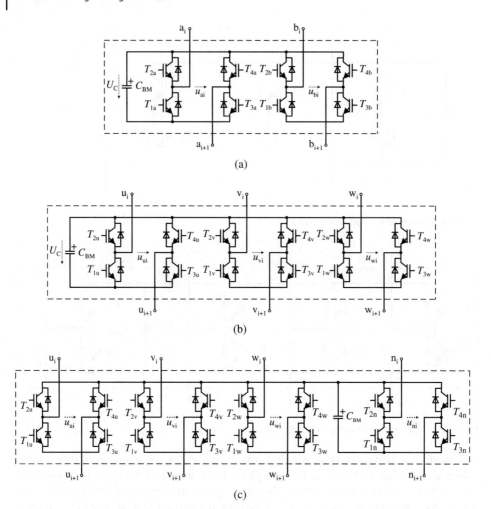

Figure 2.4 Full-bridge modules. (a) Single-phase full-bridge module. (b) Three-phase full-bridge module. (c) Three-phase four-leg full-bridge module.

(TP-FBM), and the three-phase four-leg full-bridge module (TPFL-FBM) can be obtained, as shown in Figure 2.4a–c, respectively.

Similar to the SP-HBM, the terminal voltage of the SP-FBM depends on the ON/OFF state of the switches; all statuses of the SP-FBM in Figure 2.4a are listed in Table 2.2. When T_{2a} and T_{3a} of leg a are turned ON, the terminal voltage of the full-bridge module is $u_{ai} = U_C$ and the corresponding status is named "1." Different from the SP-HBM, a negative DC voltage can be generated at the output when T_{1a} and T_{4a} of leg a are turned ON, that is $u_{ai} = -U_C$, which is named as status "−1." There are two operating modes for status "0," one is T_{1a} and T_{3a} ON and the other is T_{2a} and T_{4a} ON, which can be selected to make the switching loss evenly distributed. For the leg without a parallel capacitor (leg b), the status "0" can be obtained as well as that of leg a. However, there

Table 2.2 Switching states and terminal voltages of SP-FBM.

Leg *m*	Status	T_{1m}	T_{2m}	T_{3m}	T_{4m}	u_{mi}
	1	OFF	ON	ON	OFF	U_C
a (leg with parallel capacitor)	−1	ON	OFF	OFF	ON	$-U_C$
	0	ON	OFF	ON	OFF	0
		OFF	ON	OFF	ON	0
b (leg without parallel capacitor)	0	ON	OFF	ON	OFF	0
		OFF	ON	OFF	ON	0
	X_1	OFF	OFF	OFF	OFF	
	X_2	OFF	ON	ON	OFF	—
	X_3	ON	OFF	OFF	ON	

are three statuses referring to the condition with an uncertain terminal voltage (u_{bi}), which are "X_1," "X_2," and "X_3," respectively.

2.3 Single-Phase Half-Bridge-Module High-voltage Converter

2.3.1 Basic Structure and Operating Principle

The proposed single-phase half-bridge-module DC–AC converter is illustrated in Figure 2.5, in which both positive and negative parts consist of two leg inductors and several SP-HBMs [1]. Assume that N bridge modules are connected in series in the positive part, then terminals a_{p1} and b_{p1} of the first bridge module (SP-HBM$_{p1}$) are connected to the positive pole of the input DC voltage source U_{DC}, while terminals $a_{p(N+1)}$ and $b_{p(N+1)}$ of the last bridge module (SP-HBM$_{pN}$) are connected to the leg inductors L_{ap} and L_{bp}, respectively. As the structure of the positive and negative parts are in symmetry, the intersection of two inductors is defined as the phase output terminals, and the load is placed between phase output terminals a and b.

If the common capacitor is on leg *a*, or the bridge module of Figure 2.1a is used, based on Table 2.1, the voltage sum of the bridge modules in leg *a*, $u_{ap} = \sum_{i=1}^{N} u_{api}$ and $u_{an} = \sum_{i=1}^{N} u_{ani}$, can be controlled in the range between 0 and NU_C.

If U_C is kept at a constant value which is

$$U_C = U_{DC}/N \tag{2.1}$$

then there are N bridge modules in leg *a* which operate at status "1" at any time, because $u_{ap} + u_{an} = U_{DC}$.

When the voltage on the leg inductor is ignored, the output voltage of phase *a*, u_{ao}, can be expressed by

$$u_{ao} = \frac{U_{DC}}{2} - u_{ap} = u_{an} - \frac{U_{DC}}{2} \tag{2.2}$$

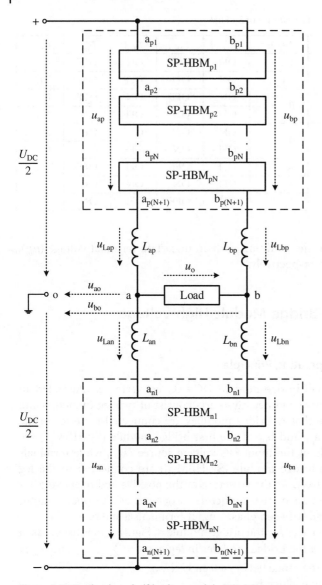

Figure 2.5 Single-phase half-bridge-module DC–AC converter.

Accordingly, u_{ao} is in multilevel form and there are N+1 voltage levels at most, which are

$$u_{ao} \in \left\{ -\frac{N}{2}U_C, -\frac{N-2}{2}U_C, \cdots, -\frac{U_C}{2}, 0, \frac{U_C}{2}, \cdots, \frac{N-2}{2}U_C, \frac{N}{2}U_C \right\} \quad (2.3)$$

It is known from Table 2.1 that the terminal voltage of the bridge module in leg *b* without a parallel capacitor is certain only at status "0." As $u_{bp} + u_{bn} = U_{DC}$ should be satisfied at any time, a simple control method is to control all the switching components on the

Table 2.3 Voltage ratios of single-phase half-bridge-module DC–AC converter.

Phase a	$\dfrac{u_{ao}}{U_C}$	$-\dfrac{N}{2}$	$\dfrac{2-N}{2}$	...	$-\dfrac{1}{2}$	0	$\dfrac{1}{2}$	...	$\dfrac{N-2}{2}$	$\dfrac{N}{2}$
Phase b	$\dfrac{u_{bo}}{U_C}$	$-\dfrac{N}{2}$	$-\dfrac{N}{2}$	...	$-\dfrac{N}{2}$	$-\dfrac{N}{2}$	$-\dfrac{N}{2}$	...	$-\dfrac{N}{2}$	$-\dfrac{N}{2}$
Output	$\dfrac{u_o}{U_C}$	0	1	...	$\dfrac{N-1}{2}$	$\dfrac{N}{2}$	$\dfrac{N+1}{2}$	...	$N-1$	N
Phase a	$\dfrac{u_{ao}}{U_C}$	$-\dfrac{N}{2}$	$\dfrac{2-N}{2}$	...	$-\dfrac{1}{2}$	0	$\dfrac{1}{2}$	...	$\dfrac{N-2}{2}$	$\dfrac{N}{2}$
Phase b	$\dfrac{u_{bo}}{U_C}$	$\dfrac{N}{2}$	$\dfrac{N}{2}$	...	$\dfrac{N}{2}$	$\dfrac{N}{2}$	$\dfrac{N}{2}$	...	$\dfrac{N}{2}$	$\dfrac{N}{2}$
Output	$\dfrac{u_o}{U_C}$	$-N$	$1-N$	...	$-\dfrac{N+1}{2}$	$-\dfrac{N}{2}$	$\dfrac{1-N}{2}$	...	1	0

positive part of leg b operating at status "0" and those on the negative part operating at status "X_2," or vice versa. Therefore, the output voltage of phase b, or u_{bo}, is expressed by

$$\begin{cases} u_{bo} = \dfrac{U_{DC}}{2} = \dfrac{NU_C}{2}, & \text{when } u_{bp} = \sum_{i=1}^{N} u_{bpi} = 0 \\[3mm] u_{bo} = -\dfrac{U_{DC}}{2} = -\dfrac{NU_C}{2}, & \text{when } u_{bn} = \sum_{i=1}^{N} u_{bni} = 0 \end{cases} \tag{2.4}$$

It should be noted that u_{bo} could be other values except $\pm\dfrac{NU_C}{2}$ when some modules on leg b are operating at status "X." In this section, u_{bo} is only controlled to be $\dfrac{NU_C}{2}$ or $-\dfrac{NU_C}{2}$, in order to make the control of the single-phase half-bridge-module DC–AC converter simple.

Based on Eqs. (2.3) and (2.4), the output voltage or load voltage $u_o = u_{ao} - u_{bo}$ can be obtained and is listed in Table 2.3. It is found that u_o is a voltage with N+1 levels, and its amplitude is NU_C or U_{DC}.

2.3.2 Control Scheme

Since the single-phase half-bridge-module DC–AC converter will consist of many bridge modules in high-voltage applications, a modulation scheme without complex calculation is desirable in practice. The traditional staircase modulation, level-shifted and phase-shifted multicarrier-based PWM schemes are popular for MMC; they are selected to implement the control of the proposed converter in this section.

Assume that n ($n \leq N$) modules in the positive part of leg a and $N-n$ modules in the negative part operate at status "1," then we have $u_{ap} = \sum_{i=1}^{N} u_{api} = nU_C$, $u_{an} = \sum_{i=1}^{N} u_{ani} = (N-n)U_C$, and $u_{ao} = \dfrac{N-2n}{2}U_C$ based on Eq. (2.2). For a specific output voltage $u_{ao} = \dfrac{N-2n}{2}U_C$, the total number of available operating modes is equal to $A_N^n \cdot A_N^{N-n}$. Take N = 2 as an example, the corresponding schematic diagram of the proposed converter is illustrated in Figure 2.6. Based on the operating principle

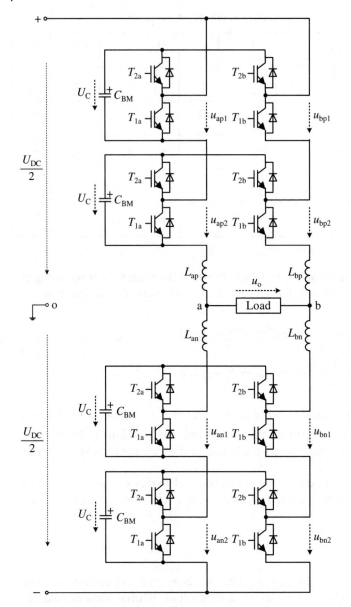

Figure 2.6 Schematic of single-phase half-bridge-module DC–AC converter when N = 2.

in Section 2.3.1, all possible operating modes of bridge modules in Figure 2.6 are summarized in Table 2.4. It is found that there are four operating modes for $u_{ao} = 0$, which is consistent with $A_2^1 \cdot A_2^1 = 4$ when $n = 1$.

Based on Table 2.4, the desired output voltage can be obtained by selecting proper operating modes. For example, if $u_o = U_{DC} = 2U_C$ is needed, then $u_{ao} = U_C$ and $u_{bo} = -U_C$ should be provided. Accordingly, the bridge modules should operate in

Table 2.4 Operating modes of single-phase half-bridge-module DC–AC converter when $N=2$.

Leg m	Mode	u_{mp1}	u_{mp2}	u_{mn1}	u_{mn2}	u_{mo}
a (with parallel capacitor)	1	0	0	U_C	U_C	U_C
	2	U_C	0	0	U_C	0
	3	U_C	0	U_C	0	0
	4	0	U_C	0	U_C	0
	5	0	U_C	U_C	0	0
	6	U_C	U_C	0	0	$-U_C$
b (without parallel capacitor)	1	0	0	U_C	U_C	U_C
	2	U_C	U_C	0	0	$-U_C$

Mode 1 for leg a and Mode 2 for leg b. As u_{bo} is a two-level waveform with $\pm U_C$ magnitudes, when $u_o = U_C$ is needed, $u_{ao} = 0$ and $u_{bo} = -U_C$ are required. At this time, Modes 2 to 5 can be selected for leg a.

Typical voltage waveforms of the proposed converter, using staircase modulation and the carrier-based PWM scheme with PODM (phase opposition disposition method), are illustrated in Figure 2.7, in which only Mode 2 of leg a is used to generate $u_{ao} = 0$.

In the staircase modulation shown in Figure 2.7a, the power devices generally switch on and off several times during one cycle of the line voltage, to minimize the switching losses in high-power applications. When the switching angles of all steps of the staircase, for example, θ_1 and θ_2, have been calculated to meet the requirements of modulation index and harmonic content, a staircase waveform following a sinusoidal envelope can be generated. In the carrier-based PWM scheme shown in Figure 2.7b, THD can be reduced at the cost of more switching losses, but this approach is quite effective when the ratio between the switching frequency and the line frequency is relatively low.

2.3.3 Output Voltage Verification

In this section, the feasibility of the proposed single-phase half-bridge-module DC–AC converter will be proven by PSIM®. The simulation prototype is the same as that in Figure 2.6, where $N=2$, $U_{DC} = 200$ V, $U_C = 100$ V, and the fundamental output frequency $f_r = 50$ Hz. The simulation waveforms of u_o, u_{ao}, u_{bo} and all bridge-module voltages using the staircase modulation scheme are shown in Figure 2.8, which are consistent with the theoretical analysis in Figure 2.7a.

The simulation waveforms of u_o, u_{ao}, u_{bo} under different frequency ratio f_s/f_r are shown in Figure 2.9, where f_s is the switching frequency. It is found that the proposed converter can output the multilevel voltage as desired.

2.3.4 Simplified Single-Phase Half-Bridge Module

Based on the control schemes described in the above section, it is found that the upper switch T_{2b} of the SP-HPM illustrated in Figure 2.1a will be kept at the OFF state all the time. Similar to a modified full-bridge module used in the MMC [2], the simplified single-phase half-bridge module (SSP-HBM) can be obtained by substituting a power

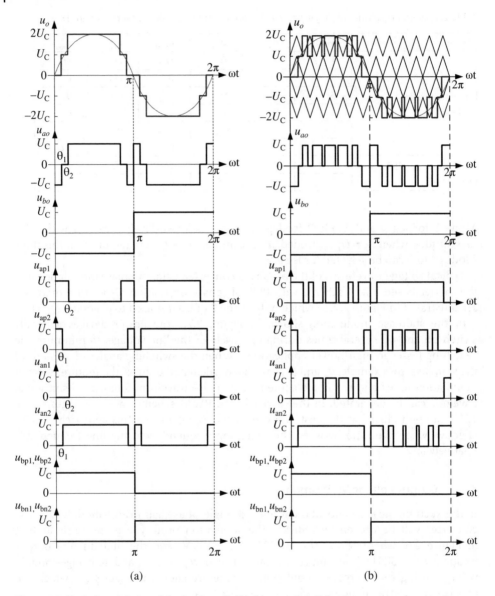

Figure 2.7 Typical waveforms of single-phase half-bridge-module DC–AC converter when N = 2. (a) Staircase modulation. (b) Carrier-based PWM.

diode D_{2b} for the power switch T_{2b}, or deleting the power switch T_{2b}. Both of the SSP-HBMs are illustrated in Figure 2.10. It should be noted that the leg with parallel capacitor is totally different from that without parallel capacitor when the SSP-HBMs are used.

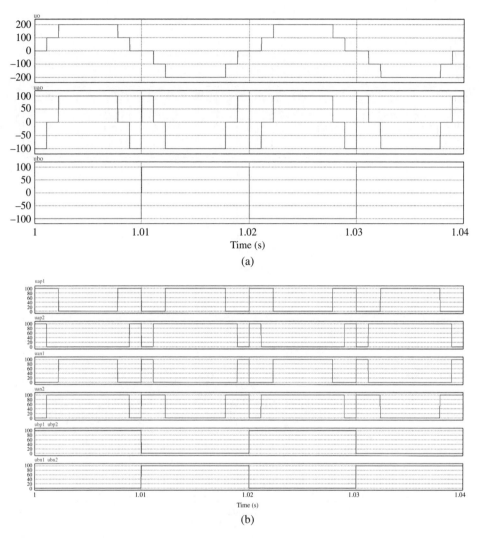

Figure 2.8 Simulation waveforms of single-phase half-bridge-module DC–AC converter with N = 2 by using staircase modulation scheme. (a) u_o, u_{ao}, and u_{bo}. (b) Terminal voltages of the bridge modules.

2.4 Three-Phase Half-Bridge-Module High-voltage Converter

2.4.1 Basic Structure and Operating Principle

Similar to the structure of the single-phase half-bridge-module DC–AC converter, the proposed three-phase half-bridge-module DC–AC converter is shown in Figure 2.11 [3], in which the line voltage is determined by the difference between phase outputs, for example, $u_{uv} = u_{uo} - u_{vo}$.

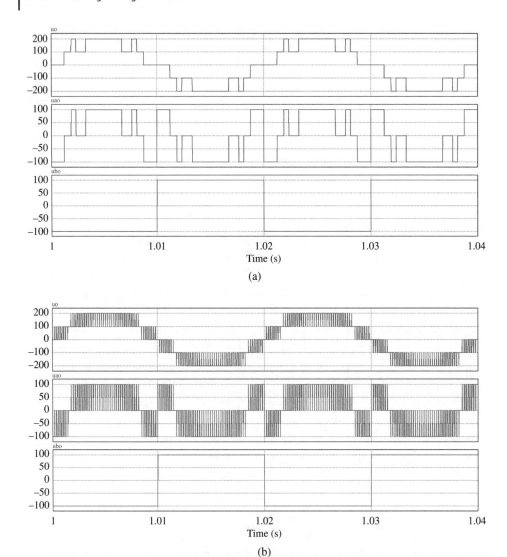

Figure 2.9 Simulation waveforms of single-phase half-bridge-module DC–AC converter with N = 2 by using carrier-based PWM scheme. (a) Frequency ratio f_s/f_r=10. (b) Frequency ratio f_s/f_r=100.

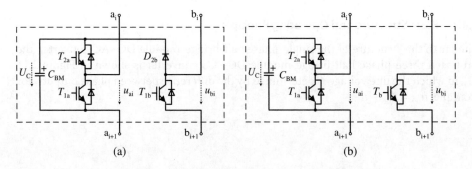

Figure 2.10 Simplified single-phase half-bridge module. (a) Type I. (b) Type II.

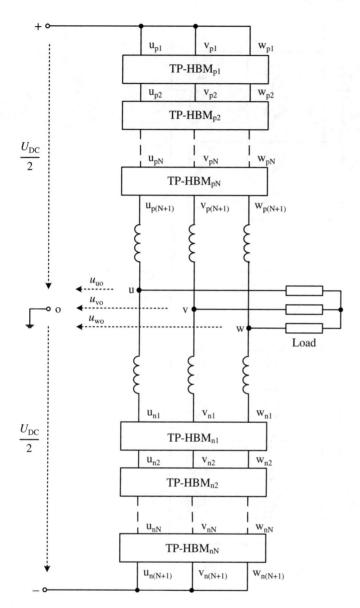

Figure 2.11 Three-phase half-bridge-module DC–AC converter.

2.4.2 Control Scheme

If the TP-HBM illustrated in Figure 2.2 is used, the common capacitor is connected on leg u, then u_{uo} can easily be controlled in the form of $N + 1$ voltage levels, as defined in Eq. (2.3). In order to obtain a symmetrical output voltage on the three-phase load, u_{vo} and u_{wo} should be multilevel, as well as u_{uo}. In Section 2.3, the phase output without parallel capacitor (u_{bo}) is controlled to have only two magnitudes, which are $\frac{U_{DC}}{2}$ and

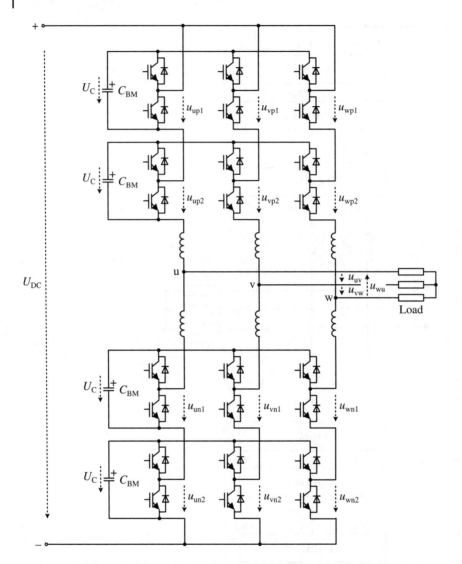

Figure 2.12 Schematic of three-phase half-bridge-module DC–AC converter with $N = 2$.

$-\frac{U_{DC}}{2}$. Obviously, this control scheme is not suitable for u_{vo} and u_{wo}. How to obtain other voltage values except $\pm\frac{NU_C}{2}$ for the phase without parallel capacitor will be a challenge for the proposed converter. In this section, the proposed three-phase half-bridge-module DC–AC converter with $N = 2$, illustrated in Figure 2.12, is taken as an example to explain the control scheme.

It is known that u_{uo} is a three-level voltage and $u_{uo} \in \{-U_C, 0, U_C\}$. Thus, $u_{vo} = 0$ and $u_{wo} = 0$ should be generated. From Figure 2.12, the following voltage relationships on leg

Table 2.5 Voltages of three-phase half-bridge-module DC–AC converter when $N = 2$.

Module	TP-HBM$_{p1}$		TP-HBM$_{p2}$		TP-HBM$_{n1}$		TP-HBM$_{n2}$		Phase
Leg m	u_{mp1}	Status	u_{mp2}	Status	u_{mn1}	Status	u_{mn2}	Status	u_{mo}
u (with parallel capacitor)	0	0	0	0	U_C	1	U_C	1	U_C
	U_C	1	0	0	0	0	U_C	1	0
	0	0	U_C	1	U_C	1	0	0	
v, w (without parallel capacitor)	U_C	1	U_C	1	0	0	0	0	$-U_C$
	0	0	0	0	U_C	X_2	U_C	X_2	U_C
	$U_C/2$	X_2	$U_C/2$	X_2	$U_C/2$	X_2	$U_C/2$	X_2	0
	U_C	X_2	U_C	X_2	0	0	0	0	$-U_C$

v are established in the steady state:

$$
\begin{cases}
u_{vo} = \dfrac{U_{DC}}{2} - (u_{vp1} + u_{vp2}) = (u_{vn1} + u_{vn2}) - \dfrac{U_{DC}}{2} \\
(u_{vp1} + u_{vp2}) + (u_{vn1} + u_{vn2}) = U_{DC}
\end{cases}
\tag{2.5}
$$

Therefore, $u_{vo} = 0$ corresponds to the operating mode with

$$
(u_{vp1} + u_{vp2}) = (u_{vn1} + u_{vn2}) = \frac{U_{DC}}{2}
\tag{2.6}
$$

Obviously, a common method to establish Eq. (2.6) is to turn all switching components on leg v OFF. Based on Eq. (2.1), $U_C = \frac{U_{DC}}{N} = \frac{U_{DC}}{2}$, then the terminal voltage of each bridge module in leg v is equal to $\frac{U_C}{2}$, that is $u_{vp1} = u_{vp2} = u_{vn1} = u_{vn2} = \frac{U_C}{2}$.

From Table 2.1, the operating status "X_2" is selected to obtain $u_{vo} = 0$, because it has the advantage that the terminals of each bridge module in leg v are not connected to other legs. By applying the above control scheme to leg w, $u_{wo} = 0$ can be obtained as well. Thus, the available voltages of the proposed three-phase half-bridge-module DC–AC converter with $N = 2$ are summarized in Table 2.5; the required phase output voltage can be created by selecting the proper statuses.

2.4.3 Output Voltage Verification

The simulation prototype is the same as that in Figure 2.12, where $N = 2$, $U_{DC} = 200$ V, $U_C = 100$ V, and the fundamental output frequency $f_r = 50$ Hz. By applying the staircase modulation, the simulation waveforms of phase voltages, line voltages, and some terminal voltages are shown in Figure 2.13. As shown in Figure 2.13a,b, both u_{uo} and u_{vo} are the three-level voltage as desired, which proves the feasibility of the control scheme in Table 2.5. Figure 2.13c illustrates the simulation results of three phase outputs and three line voltages; it is found that the proposed converter can output the three-phase multilevel voltages as well as the traditional MMC.

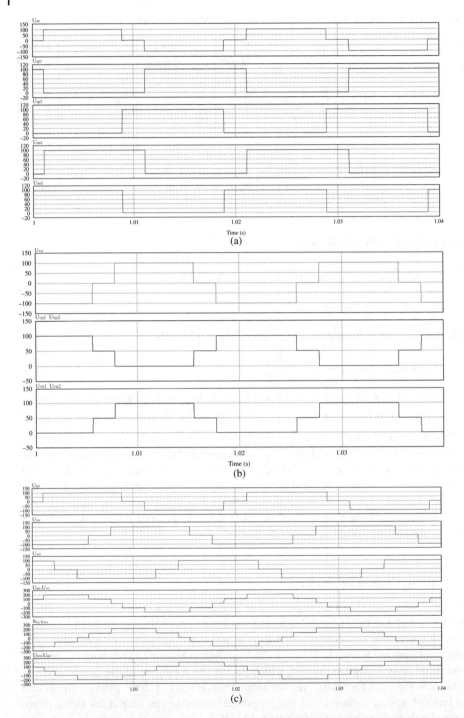

Figure 2.13 Simulation waveforms of three-phase half-bridge-module DC–AC converter (N = 2) by using staircase modulation scheme. (a) u_{uo}, u_{up1}, u_{up2}, u_{un1}, and u_{un2}. (b) u_{vo}, u_{vp1}, u_{vp2}, u_{vn1}, and u_{vn2}. (c) Phase voltages and line voltages.

For the proposed three-phase half-bridge-module DC–AC converter, when $N > 2$ it is predicted that the control schemes to generate the voltage magnitudes $\pm\frac{kU_C}{2}$ (where $k = 1, 2, \ldots, N-1$) will be more complicated than those for 0 and $\pm\frac{NU_C}{2}$. At this time, the status "X_1" can be utilized to produce the desired voltage for the leg without parallel capacitor.

2.5 Three-Phase Four-Leg Half-Bridge-Module High-voltage Converter

By applying the three-phase four-leg half-bridge module in Figure 2.3, the proposed high-voltage DC–AC converter is illustrated in Figure 2.14 [4], in which the load voltage of each phase is determined by $u_{xn} = u_{xo} - u_{no}$, where $x = u, v, w$.

Since the capacitor is in parallel with the neutral leg in the TPFL-HBM, the corresponding phase output u_{no} is in the form of N + 1 levels: $\left\{-\frac{N}{2}U_C, -\frac{N-2}{2}U_C, \cdots, -\frac{U_C}{2}, 0, \frac{U_C}{2}, \cdots, \frac{N-2}{2}U_C, \frac{N}{2}U_C\right\}$. By using the control scheme described in Section 2.4.2, the three-phase outputs u_{uo}, u_{vo}, u_{wo} can be controlled to be three-level voltages, whose magnitudes are $-\frac{NU_C}{2}$, 0, $\frac{NU_C}{2}$. Therefore, the symmetrical phase voltages can be obtained, and the specific control scheme for u_{no} can be applied to solve the load unbalance problem, harmonic compensation, and so on.

2.6 Full-Bridge-Module High-voltage Converter

By replacing the half-bridge modules in Figures 2.5, 2.11, and 2.14 with the full-bridge ones, full-bridge-module high-voltage converters can be built [5]. Comparing the statuses of the full-bridge module (Table 2.2) with those of the half-bridge module (Table 2.1), the only difference between them is that there is an extra status "−1" for the leg with parallel capacitor in the full-bridge module. Therefore, by applying the control schemes for the half-bridge-module DC–AC converter to the full-bridge-module DC–AC converter, the full-bridge-module DC–AC converter will output the multilevel voltage as well as the half-bridge-module DC–AC converter.

Since the full-bridge module can operate in status "−1," the voltage sum of bridge modules in the leg with parallel capacitor can be controlled in the range between $-NU_C$ and NU_C. When the SP-FBM in Figure 2.3 is used, the following equation is established if $U_C = \frac{U_{DC}}{N}$:

$$-NU_C \leq u_{ap},\ u_{an} \leq NU_C \text{ or } -U_{DC} \leq u_{ap},\ u_{an} \leq U_{DC} \tag{2.7}$$

For the full-bridge-module high-voltage converter with N = 2, the voltages in phase *a* have the following relationship when the leg inductor voltages are considered:

$$u_{ap} + u_{Lap} + u_{an} + u_{Lan} = U_{DC} \tag{2.8}$$

Then the output voltage of phase *a*, u_{ao}, is expressed by

$$u_{ao} = \frac{U_{DC}}{2} - u_{ap} - u_{Lap} = u_{an} + u_{Lan} - \frac{U_{DC}}{2} \tag{2.9}$$

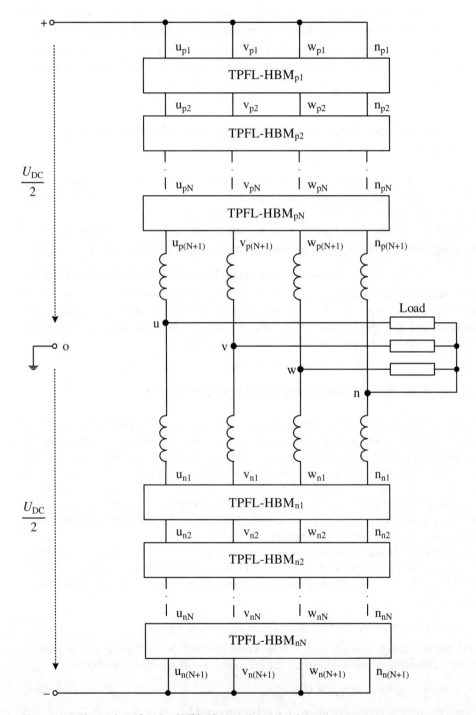

Figure 2.14 Three-phase four-leg half-bridge-module DC–AC converter.

It is obvious that the peak value of u_{ao} can be a little larger than $\frac{U_{DC}}{2}$, which can reduce the input current to the converter and the total loss of the converter.

2.7   Advantages of Multiple-Bridge-Module Converter

In this section, the proposed MBMC will be compared with the conventional MMC in terms of connection, component number, and power loss. First, each phase of the MMC is independent and the structure of each leg is the same, in which N sub-modules are connected in series. But in the MBMC, the upper or lower legs of different phases are integrated together; it is bridge modules which are connected in series. Thus, the "modulability" or power density of the MBMC will be higher than that of the MMC.

In order to compare the numbers of components used in the converters, the three-phase half-bridge-module converter is taken as example, while the sub-module of the MMC is in the form of a half-bridge. The comparison results are listed in Table 2.6 when the series number of modules is N.

It is found that the numbers of power switches and inductors used in the MBMC are equal to those used in the MMC, while the number of capacitors used in the MBMC is only one-third of that used in the MMC. As a result, the cost and volume of the MBMC can be reduced to a lower level.

As we know, the power loss of a power switch, including conduction loss and switching loss, is the main factor affecting the efficiency of high-voltage converters. In the MMC, two power switches in the half-bridge sub-module (HBSM) are turned ON or OFF complementarily, which means that there is one switch in every half-bridge sub-module operating in the ON state at any time. Thus, the total number of "ON" switches is 6N.

Based on the operating principle of a three-phase half-bridge-module converter (Table 2.5), one power switch on the leg with parallel capacitor in every TP-HBM must be turned ON at any time. However, both power switches on the leg without parallel capacitor are turned OFF when status "X_2" is selected, while only the lower switch is turned ON when status "0" is selected. Therefore, the total number of "ON" switches is 2N when all the legs without parallel capacitor are operating at status "X_2", and 4N when they are operating at status "0." Assuming that the switching times in one switching period are equal, the total power loss of the power switches in the MBMC is obviously less than that of the MMC.

Though the proposed MBMC has several advantages over the MMC in terms of power density, capacitor number, and power loss, the control scheme for legs with parallel capacitor in the MBMC is different from that for legs without parallel capacitor, which will make the control methods for the MBMC a little complicated.

Table 2.6 Comparison between MMC and MBMC.

Converter	Type of module	Number of components		
		Power switch	Capacitor	Inductor
MMC	HBSM	12N	6N	6
MBMC	TP-HBM	12N	2N	6

2.8 Summary

A kind of novel multiple-bridge-module DC–AC converter has been put forward for the first time in this chapter, which has advantages of higher power density, fewer number of capacitor, lower power loss, and reduced cost compared with the traditional MMC. It is expected that the proposed MBMC will provide an optional solution with fewer capacitors than the MMC in high-voltage applications.

References

1 Zhang, B., Qiu, D. Y. Single-phase high-voltage converter with H-bridge modules and its multilevel control method. State Intellectual Property Office of the P.R.C., ZL 201310719815.0, 2017.4.12.
2 Zhang, B., Li, X. F., Qiu, D. Y., Zhang, G. D. A kind of sub-module circuit for modular multilevel converter. State Intellectual Property Office of the P.R.C., ZL 201510222531.X, 2017.10.20.
3 Zhang, B., Qiu, D. Y. Three-phase high-voltage converter with bridge modules. State Intellectual Property Office of the P.R.C., ZL 201310719758.6, 2017.1.11.
4 Zhang, B., Qiu, D. Y. Three-phase four-wire high-voltage converter with bridge modules. State Intellectual Property Office of the P.R.C., ZL 2014100733394, 2016.8.17.
5 Zhang, B., Qiu, D. Y. Three-phase high-voltage converter with H-bridge modules. State Intellectual Property Office of the P.R.C., ZL 201310719788.7, 2016.10.5.

3

Single-Input Multiple-Output High-voltage DC–AC Converters

3.1 Introduction

In order to satisfy the requirement of multiple AC outputs in some high-voltage high-power applications, several novel single-phase/three-phase DC–AC converters with two or more outputs are proposed in this chapter. The circuit topology and operating principle are described in detail, the carrier phase-shifted sinusoidal pulse-width modulation (CPS-SPWM) scheme is applied to the proposed converters, and the simulation waveforms are provided to prove the feasibility of the proposed multiple-output high-voltage DC–AC converters.

3.2 Single-Input Dual-Output Half-Bridge Single-Phase DC–AC Converter

3.2.1 Basic Structure and Operating Principle

The proposed single-input dual-output half-bridge single-phase DC–AC converter is shown in Figure 3.1, composed of a DC voltage source, three input capacitors (C_1, C_2, and C_3), three switching arms, and one coupled inductor [1]. The input capacitors are connected in series and then paralleled with the DC voltage source. The upper switching arm (U), the primary winding of the coupled inductor (L_p), the middle switching arm (M), the secondary winding of the coupled inductor (L_s), and the lower switching arm (L) are connected in turn. Therefore, load #1 is connected to the endpoint a_1 of the middle switching arm (M) and the intersection point o_1 of capacitors C_1 and C_2, and load #2 is connected to another endpoint a_2 of the middle switching arm (M) and the intersection point o_2 of capacitors C_2 and C_3.

Each switching arm consists of N sub-modules (SMs), which are connected in series. There are different kinds of SM, and the half-bridge topology shown in Figure 3.2 is the common one, made up of one capacitor C_{SM}, two full-controlled power switches (T_1, T_2), and their antiparallel diodes (D_1, D_2).

The SM could be considered as a controlled voltage source, because its output voltage depends on the ON or OFF state of the power switches. When the upper switch T_2 is turned ON and the lower switch T_1 is turned OFF, the capacitor C_{SM} is inserted into the switching arm and the output voltage of the SM, u_{SM}, is equal to the capacitor voltage U_C, that is, $u_{SM} = U_C$. When T_1 is turned ON and T_2 is turned OFF, the capacitor C_{SM}

Multi-terminal High-voltage Converter, First Edition. Bo Zhang and Dongyuan Qiu.
© 2019 John Wiley & Sons Singapore Pte. Ltd. Published 2019 by John Wiley & Sons Singapore Pte. Ltd.

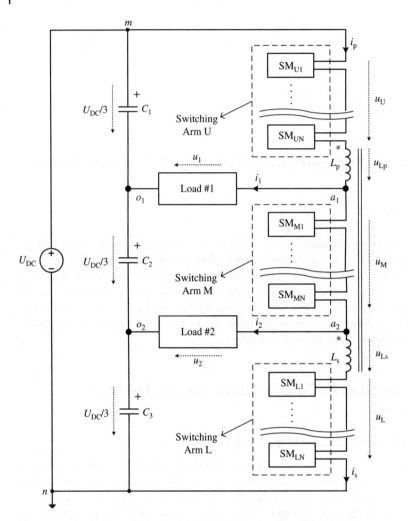

Figure 3.1 Single-input dual-output half-bridge single-phase DC–AC converter.

Figure 3.2 Half-bridge sub-module.

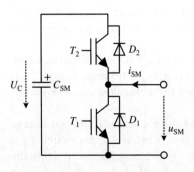

is bypassed and the output voltage u_{SM} becomes zero, that is, $u_{SM} = 0$. Therefore, the output voltage of the whole switching arm is the sum of the output voltages of N SMs within the arm, that is, $u_X = \sum_{i=1}^{N} u_{SMXi}$, where X = U, M, L.

Assuming that all components are ideal, the primary inductance L_p, secondary inductance L_s, and mutual inductance M_L of the coupled inductor are equal, that is, $L_p = L_s = M_L$. As $u_{Lp} = L_p \frac{di_p}{dt} + M_L \frac{di_s}{dt}$ and $u_{Ls} = L_s \frac{di_s}{dt} + M_L \frac{di_p}{dt}$, the following voltage relationship can be obtained:

$$u_{Lp} = u_{Ls} \tag{3.1}$$

Therefore,

$$u_{a1n} + u_{a2n} = U_{DC} - u_U + u_L \tag{3.2}$$

From Figure 3.1, we have

$$u_{a1n} - u_{a2n} = u_M \tag{3.3}$$

where U_{DC} is the input DC voltage and u_U, u_M, u_L are the output voltages of the upper switching arm (U), middle switching arm (M), and lower switching arm (L), respectively.

Based on Eqs. (3.2) and (3.3), the voltages at the intersection points a_1 and a_2 are expressed by

$$\begin{cases} u_{a2n} = \dfrac{u_L + u_M - u_U}{2} + \dfrac{U_{DC}}{2} \\ u_{a2n} = \dfrac{u_L - u_M - u_U}{2} + \dfrac{U_{DC}}{2} \end{cases} \tag{3.4}$$

Then, the dual output voltages in Figure 3.1 are determined by

$$\begin{cases} u_1 = u_{a1n} - \dfrac{2}{3} U_{DC} = \dfrac{U_{DC}}{2} \left(\dfrac{u_L + u_M - u_U}{U_{DC}} - \dfrac{1}{3} \right) \\ u_2 = u_{a2n} - \dfrac{1}{3} U_{DC} = \dfrac{U_{DC}}{2} \left(\dfrac{u_L - u_M - u_U}{U_{DC}} + \dfrac{1}{3} \right) \end{cases} \tag{3.5}$$

3.2.2 Control Scheme

In this section, the CPS-SPWM scheme is applied to the proposed single-input dual-output half-bridge single-phase DC–AC inverter. If there are N SMs in each switching arm, then N triangular carrier signals phase shifted by an angle 360°/N are needed.

Normally, the voltages across the loads are expected to be sinusoidal. Based on Eq. (3.5), if the reference signals to regulate u_{a1n} and u_{a2n} are defined by the following equations:

$$\begin{cases} u_{Ref1}(t) = \dfrac{u_L + u_M - u_U}{U_{DC}} = M_1 \sin(2\pi f_1 t + \varphi_1) + \dfrac{1}{3} \\ u_{Ref2}(t) = \dfrac{u_L - u_M - u_U}{U_{DC}} = M_2 \sin(2\pi f_2 t + \varphi_2) - \dfrac{1}{3} \end{cases} \tag{3.6}$$

then the output voltages of the proposed half-bridge converter will be

$$\begin{cases} u_1(t) = \dfrac{U_{DC}}{2} M_1 \sin(2\pi f_1 t + \varphi_1) \\ u_2(t) = \dfrac{U_{DC}}{2} M_2 \sin(2\pi f_2 t + \varphi_2) \end{cases} \qquad (3.7)$$

where M_1 and M_2 are modulated ratios, f_1 and f_2 are frequencies, φ_1 and φ_2 are phase shifts. If the modulation space of the CPS-SPWM scheme is set to be $[-1, 1]$, and the peak-to-peak value of the carried waves is 2, then $M_1, M_2 \leq \frac{2}{3}$ to avoid over-modulation.

It is clear from Eq. (3.6) that there are two operating modes for the proposed converter according to the frequencies of the output voltage, one is the different frequency (DF) mode and the other is the equal frequency (EF) mode. Therefore, the relationship among two reference signals u_{Ref1}, u_{Ref2} and the carrier signals under the DF mode is shown in Figure 3.3a, while that under the EF mode is shown in Figure 3.3b, and Figure 3.3c refers to the EF case when both of the reference signals are in phase, which is named the EF′ mode.

In order to avoid overlap of output voltages, $u_{Ref1}(t) > u_{Ref2}(t)$ is required and the modulating ratios have the following relationships:

$$\begin{cases} M_1 + M_2 \leq \dfrac{2}{3} & \text{for DF and EF modes} \\ M_1, M_2 \leq \dfrac{2}{3} & \text{for EF′ mode} \end{cases} \qquad (3.8)$$

According to Eq. (3.6), the switching-arm voltages u_L, u_M, and u_U should be controlled to satisfy the following equations:

$$\begin{cases} -u_U + u_L = \dfrac{U_{DC}}{2}(u_{Ref1} + u_{Ref2}) \\ u_M = \dfrac{U_{DC}}{2}(u_{Ref1} - u_{Ref2}) \end{cases} \qquad (3.9)$$

Thus, one of the control strategies is generating u_U and u_L by u_{Ref1} and u_{Ref2}, respectively, and u_M by combining the control signals for u_U and u_L.

Assume that the control signal is used to control the power switch T_1 in the corresponding SM, while T_2 is controlled by the complementary signal to avoid the capacitor C_{SM} being short-circuited. Thus, when the control signal is equal to 1, T_1 will be turned ON and T_2 turned OFF; the output voltage of the SM will be $u_{SM} = 0$ based on Figure 3.2.

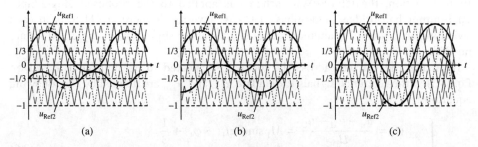

(a)   (b)   (c)

Figure 3.3 Relationship between carrier signals and reference signals of the single-input dual-output half-bridge single-phase DC–AC converter. (a) DF mode. (b) EF mode. (c) EF′ mode.

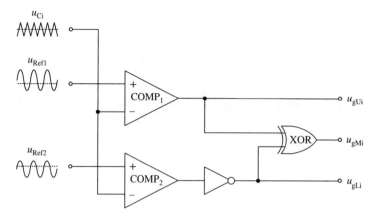

Figure 3.4 Control schematic of the single-input dual-output half-bridge single-phase DC–AC converter.

Otherwise, when the control signal is equal to 0, T_1 will be turned OFF and T_2 turned ON, which results in $u_{SM} = U_C$.

Based on the above SM control method and the CPS-SPWM theory, the control signals for the ith SM in the upper, middle, and lower switching arms can be obtained by comparing the ith carrier signal u_{Ci} (where $i = 1, 2, ..., N$) to the reference signals u_{Ref1} and u_{Ref2}, as shown in Figure 3.4. The gate signal for the ith upper SM, u_{gUi}, are the positive logic values generated by u_{Ci} and u_{Ref1}, the gate signal for the lower SM, u_{gLi}, are the negative logic values generated by u_{Ci} and u_{Ref2}, and the gate signal for the middle SM, u_{gMi}, are the logical XOR values of u_{gUi} and u_{gLi}, that is, $u_{gMi} = u_{gUi} \oplus u_{gLi}$.

From Figure 3.4, $u_{gUi} = 1$ when $u_{Ref1} > u_{Ci}$, however $u_{gLi} = 1$ when $u_{Ref2} < u_{Ci}$. And $u_{gMi} = 1$ if either of the XOR gate inputs is equal to 1. As $u_{Ref1}(t) > u_{Ref2}(t)$, there are only three kinds of value combination for u_{gUi}, u_{gMi}, or u_{gLi}, which are 110, 101, and 011, then only one SM among SM_{Ui}, SM_{Mi}, and SM_{Li} will output voltage U_C. Therefore, there are N SMs with output voltage U_C and $2N$ SMs with zero output voltage at any moment. When the capacitor voltage of the SM is controlled to be $U_C = \frac{U_{DC}}{N}$, the output voltages of the three switching arms will be $u_L + u_M + u_U = U_{DC}$.

3.2.3 Output Voltage Verification

When $N = 3$, $U_{DC} = 300\,\text{V}$, 10Ω and $0.1\,\text{H}$ are selected as the load impendence for both outputs, the proposed single-input dual-output half-bridge single-phase DC–AC converter will operate in EF mode (in phase) if $u_{Ref1} = \frac{2}{3} \sin 100\pi t + \frac{1}{3}$ and $u_{Ref2} = \frac{1}{3} \sin 100\pi t - \frac{1}{3}$. The simulation waveforms of the reference signals u_{Ref1} and u_{Ref2}, the first output voltage u_1, the second output voltage u_2, and the two output currents i_1 and i_2 are shown in Figure 3.5a. It is shown that both output voltages/currents have the same frequency and phase angle. When $u_{Ref1} = \frac{1}{3} \sin 200\pi t + \frac{1}{3}$ and $u_{Ref2} = \frac{1}{4} \sin\left(100\pi t + \frac{\pi}{2}\right) - \frac{1}{3}$, the corresponding simulation waveforms are shown in Figure 3.5b, which prove that the proposed converter is operating in DF mode. Thus, the proposed converter can convert the single DC input voltage to dual AC output voltages with identical or DF.

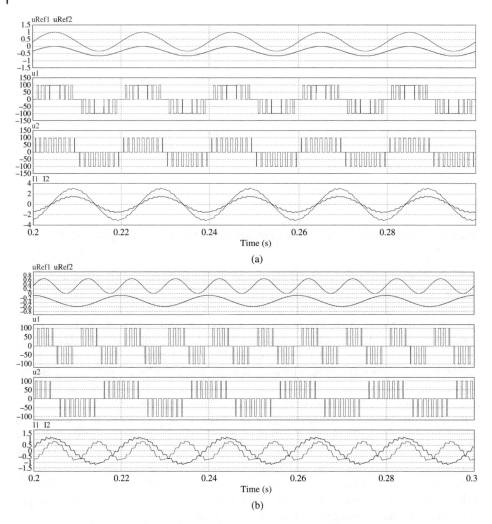

Figure 3.5 Simulation waveforms of the single-input dual-output half-bridge single-phase DC–AC converter. (a) EF mode. (b) DF mode.

3.3 Single-Input Dual-Output Full-Bridge Single-Phase DC–AC Converter

3.3.1 Basic Structure and Operating Principle

The single-input dual-output full-bridge single-phase DC–AC converter can be developed by replacing the three series capacitors in Figure 3.1 with three switching arms and one coupled inductor, as shown in Figure 3.6 [2]. Considering the three switching arms and one coupled inductor as one phase unit, there are two phase units *a* and *b* in the proposed full-bridge single-phase DC–AC converter, and the output loads are connected between two phase units.

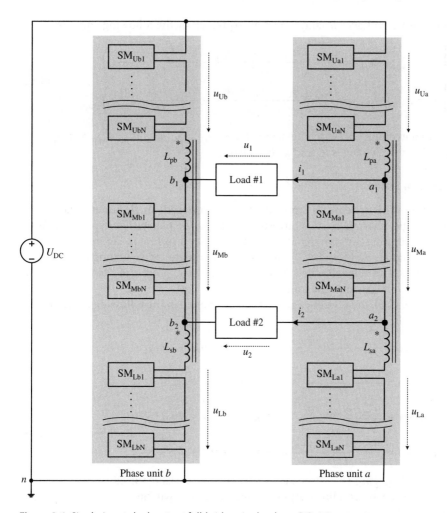

Figure 3.6 Single-input dual-output full-bridge single-phase DC–AC converter.

Similar to Eq. (3.4), the voltage relationships of phase units *a* and *b* in Figure 3.6 can be defined by

$$
\begin{cases}
u_{\text{a1n}} = \dfrac{u_{\text{La}} + u_{\text{Ma}} - u_{\text{Ua}}}{2} + \dfrac{U_{\text{DC}}}{2} \\[2mm]
u_{\text{a2n}} = \dfrac{u_{\text{La}} - u_{\text{Ma}} - u_{\text{Ua}}}{2} + \dfrac{U_{\text{DC}}}{2}
\end{cases}
\tag{3.10}
$$

$$
\begin{cases}
u_{\text{b1n}} = \dfrac{u_{\text{Lb}} + u_{\text{Mb}} - u_{\text{Ub}}}{2} + \dfrac{U_{\text{DC}}}{2} \\[2mm]
u_{\text{b2n}} = \dfrac{u_{\text{Lb}} - u_{\text{Mb}} - u_{\text{Ub}}}{2} + \dfrac{U_{\text{DC}}}{2}
\end{cases}
\tag{3.11}
$$

The dual-output voltages u_1 and u_2 can be expressed by the following equations:

$$\begin{cases} u_1 = u_{a1n} - u_{b1n} = \dfrac{U_{DC}}{2}\left(\dfrac{u_{La} + u_{Ma} - u_{Ua}}{U_{DC}} - \dfrac{u_{Lb} + u_{Mb} - u_{Ub}}{U_{DC}}\right) \\[2mm] u_2 = u_{a2n} - u_{b2n} = \dfrac{U_{DC}}{2}\left(\dfrac{u_{La} - u_{Ma} - u_{Ua}}{U_{DC}} - \dfrac{u_{Lb} - u_{Mb} - u_{Ub}}{U_{DC}}\right) \end{cases} \tag{3.12}$$

Thus, by controlling six switching arm voltages u_{Ua}, u_{Ma}, u_{La}, u_{Ub}, u_{Mb}, and u_{Lb}, two different AC output voltages u_1 and u_2 can be obtained.

3.3.2 Control Scheme

If the CPS-SPWM scheme is applied to the proposed single-input dual-output full-bridge single-phase DC–AC converter as well as the half-bridge type, the control scheme for generating the gate signals for SMs is the same as that in Figure 3.4, in which the reference signals u_{Refa1} and u_{Refa2} are used to modulate the switching-arm voltages in phase unit a, and the reference signals u_{Refb1} and u_{Refb2} are used to modulate the switching-arm voltages in phase unit b.

Similar to Eq. (3.6), the reference signals can be defined in the following forms to simplify the control:

$$\begin{cases} u_{Refa1}(t) = \dfrac{u_{La} + u_{Ma} - u_{Ua}}{U_{DC}} = M_1 \sin(2\pi f_1 t + \varphi_1) + U_{os1} \\[2mm] u_{Refb1}(t) = \dfrac{u_{Lb} + u_{Mb} - u_{Ub}}{U_{DC}} = -M_1 \sin(2\pi f_1 t + \varphi_1) + U_{os1} \end{cases} \tag{3.13}$$

$$\begin{cases} u_{Refa2}(t) = \dfrac{u_{La} - u_{Ma} - u_{Ua}}{U_{DC}} = M_2 \sin(2\pi f_2 t + \varphi_2) + U_{os2} \\[2mm] u_{Refb2}(t) = \dfrac{u_{Lb} - u_{Mb} - u_{Ub}}{U_{DC}} = -M_2 \sin(2\pi f_2 t + \varphi_2) + U_{os2} \end{cases} \tag{3.14}$$

where U_{os1} and U_{os2} are DC offsets, $U_{os1} > 0$ and $U_{os2} < 0$ as required.

Therefore, the corresponding output voltages u_1 and u_2 will be

$$\begin{cases} u_1(t) = \dfrac{U_{DC}}{2}(u_{refa1}(t) - u_{refb1}(t)) = U_{DC}{\cdot}M_1 \sin(2\pi f_1 t + \varphi_1) \\[2mm] u_2(t) = \dfrac{U_{DC}}{2}(u_{refa2}(t) - u_{refb2}(t)) = U_{DC}{\cdot}M_2 \sin(2\pi f_2 t + \varphi_2) \end{cases} \tag{3.15}$$

It is obvious that the DC voltage utilization of the full-bridge type is double that of the half-bridge type.

According to Eqs. (3.13) and (3.14), the relationships among the reference signals and the carrier signals of the proposed full-bridge converter in both DF and EF' modes are shown in Figure 3.7. In order to fully utilize the DC input voltage, $U_{os1} = 1 - M_1$ and $U_{os2} = M_2 - 1$, the modulating ratios of two reference signals should satisfy $M_1 + M_2 \le 1$ in DF and EF modes, while $|M_1 - M_2| \le U_{os2} - U_{os1}$ in EF' mode.

3.3.3 Output Voltage Verification

The simulation parameters are $N = 4$, $U_{DC} = 200\,\text{V}$, $L_{AC} = 0.1\,\text{H}$, and $R_{AC} = 10\,\Omega$ for both output loads. The proposed single-input dual-output full-bridge single-phase

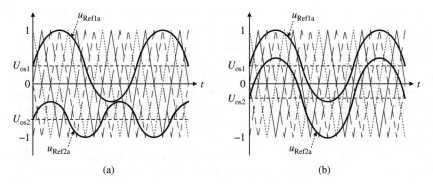

Figure 3.7 Relationship between carrier signals and reference signals of the single-input dual-output full-bridge single-phase DC–AC converter. (a) DF mode. (b) EF′ mode.

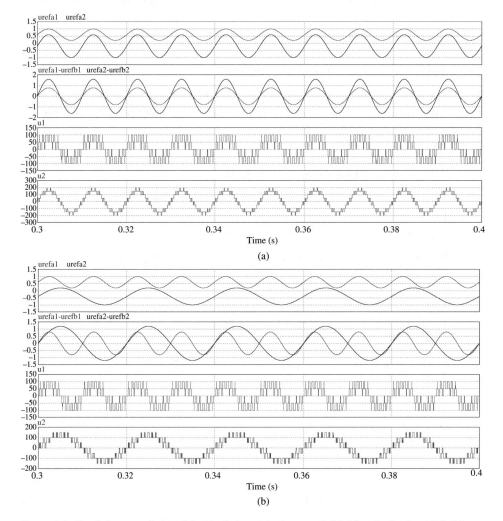

Figure 3.8 Simulation waveforms of the single-input dual-output full-bridge single-phase DC–AC converter. (a) EF′ mode. (b) DF mode.

DC–AC converter will operate in EF′ mode if $u_{Refa1} = 0.4 \sin 200\pi t + 0.6$ and $u_{Refa2} = 0.8 \sin 200\pi t - 0.2$. The simulation waveforms of the reference signals u_{Refa1} and u_{Refa2}, the modulating signals $u_{Refa1} - u_{Refb1}$ for output voltage u_1 and $u_{Refa2} - u_{Refb2}$ for output voltage u_2, and the first output voltage u_1 and the second output voltage u_2 are shown in Figure 3.8a. It is shown that both output AC voltages/currents have the same frequency and phase angle, but with different amplitude. When $u_{Refa1} = 0.4 \sin 200\pi t + 0.6$ and $u_{Refa2} = 0.6 \sin 100\pi t - 0.4$ are selected, the corresponding simulation waveforms are shown in Figure 3.8b, which proves that the proposed converter is operating in DF mode.

From the above analysis, the proposed single-input dual-output full-bridge single-phase DC–AC converter can output dual AC output signals with identical or DF as well as the half-bridge one. Though more switching components are needed in the proposed full-bridge high-voltage converter, the DC voltage utilization is double and the output voltage level is improved, which makes the output signal quality better.

3.4 Single-Input Dual-Output Three-Phase DC–AC Converter

3.4.1 Basic Structure and Operating Principle

By adding one more phase unit to the converter in Figure 3.6, a novel single-input dual-output three-phase DC–AC converter is obtained in Figure 3.9, which consists of three phase units, u, v, and w [3, 4].

Assume that the coupled inductances and the corresponding mutual inductance of each phase unit are identical, that is $L_{pj} = L_{sj} = M_j$, where $j = u$, v, w. Similar to Eq. (3.9), the dual three-phase output voltages are defined by

$$\begin{cases} u_{uv1} = u_{u1} - u_{v1} = \dfrac{u_{Lu} + u_{Mu} - u_{Uu}}{2} - \dfrac{u_{Lv} + u_{Mv} - u_{Uv}}{2} \\[2mm] u_{vw1} = u_{v1} - u_{w1} = \dfrac{u_{Lv} + u_{Mv} - u_{Uv}}{2} - \dfrac{u_{Lw} + u_{Mw} - u_{Uw}}{2} \\[2mm] u_{wu1} = u_{w1} - u_{u1} = \dfrac{u_{Lw} + u_{Mw} - u_{Uw}}{2} - \dfrac{u_{Lu} + u_{Mu} - u_{Uu}}{2} \end{cases} \tag{3.16}$$

$$\begin{cases} u_{uv2} = u_{u2} - u_{v2} = \dfrac{u_{Lu} - u_{Mu} - u_{Uu}}{2} - \dfrac{u_{Lv} - u_{Mv} - u_{Uv}}{2} \\[2mm] u_{vw2} = u_{v2} - u_{w2} = \dfrac{u_{Lv} - u_{Mv} - u_{Uv}}{2} - \dfrac{u_{Lw} - u_{Mw} - u_{Uw}}{2} \\[2mm] u_{wu2} = u_{w2} - u_{u2} = \dfrac{u_{Lw} - u_{Mw} - u_{Uw}}{2} - \dfrac{u_{Lu} - u_{Mu} - u_{Uu}}{2} \end{cases} \tag{3.17}$$

3.4.2 Control Scheme

The CPS-SPWM scheme presented in Section 3.2.2 can be applied to the single-input dual-output three-phase DC–AC converter, while the three-phase modulating references u_{Refj1} and u_{Refj2} are defined by

$$u_{Refj1}(t) = M_1 \sin(2\pi f_1 t + \varphi_1 + \phi_j) + U_{os1} \tag{3.18}$$

$$u_{Refj2}(t) = M_2 \sin(2\pi f_2 t + \varphi_2 + \phi_j) + U_{os2} \tag{3.19}$$

where $\phi_u = 0$, $\phi_v = -\frac{2\pi}{3}$, $\phi_w = \frac{2\pi}{3}$, and Eq. (3.15) should be obeyed as well.

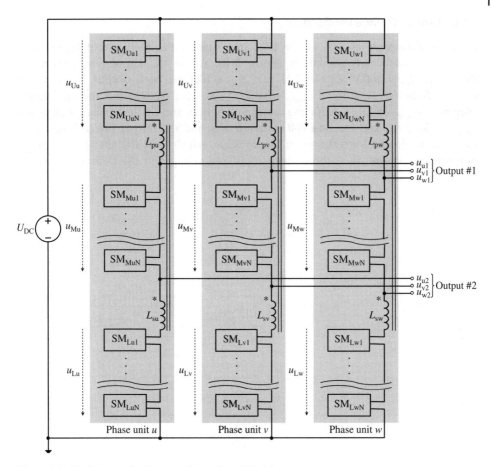

Figure 3.9 Single-input dual-output three-phase DC–AC converter.

As a result, the corresponding three-phase output voltages will be

$$
\begin{cases}
u_{uv1}(t) = \dfrac{U_{DC}}{2}(u_{refu1}(t) - u_{refv1}(t)) = \dfrac{\sqrt{3}}{2}U_{DC}{\cdot}M_1 \sin\left(2\pi f_1 t + \varphi_1 - \dfrac{\pi}{3}\right) \\[2mm]
u_{vw1}(t) = \dfrac{U_{DC}}{2}(u_{refv1}(t) - u_{refw1}(t)) = \dfrac{\sqrt{3}}{2}U_{DC}{\cdot}M_1 \sin(2\pi f_1 t + \varphi_1 - \pi) \\[2mm]
u_{wu1}(t) = \dfrac{U_{DC}}{2}(u_{refw1}(t) - u_{refu1}(t)) = \dfrac{\sqrt{3}}{2}U_{DC}{\cdot}M_1 \sin\left(2\pi f_1 t + \varphi_1 + \dfrac{\pi}{3}\right)
\end{cases}
$$
(3.20)

$$
\begin{cases}
u_{uv2}(t) = \dfrac{U_{DC}}{2}(u_{refu2}(t) - u_{refv2}(t)) = \dfrac{\sqrt{3}}{2}U_{DC}{\cdot}M_2 \sin\left(2\pi f_2 t + \varphi_2 - \dfrac{\pi}{3}\right) \\[2mm]
u_{vw2}(t) = \dfrac{U_{DC}}{2}(u_{refv2}(t) - u_{refw2}(t)) = \dfrac{\sqrt{3}}{2}U_{DC}{\cdot}M_2 \sin(2\pi f_2 t + \varphi_2 - \pi) \\[2mm]
u_{wu2}(t) = \dfrac{U_{DC}}{2}(u_{refw2}(t) - u_{refu2}(t)) = \dfrac{\sqrt{3}}{2}U_{DC}{\cdot}M_2 \sin\left(2\pi f_2 t + \varphi_2 + \dfrac{\pi}{3}\right)
\end{cases}
$$
(3.21)

3.4.3 Output Voltage Verification

Some simulation waveforms of the proposed single-input dual-output three-phase DC–AC converter are shown in Figure 3.10, where the simulation parameters are $N = 4$, $U_{DC} = 200$ V, $L_{AC} = 0.1$ H, and $R_{AC} = 10\Omega$ for both output loads. When $u_{Refu1} = 0.5 \sin 100\pi t + 0.5$ and $u_{Refu2} = 0.8 \sin 100\pi t - 0.2$ are selected, the modulating references u_{Refu1} and u_{Refu2}, the line voltage references and the line voltages (u_{uv1} and u_{vw1}) for load #1, and the line voltage references and the output line voltages (u_{uv2} and u_{vw2}) for load #2 are shown in Figure 3.10a, which indicates that the proposed converter can output dual three-phase multilevel voltages in the desired EF′ mode.

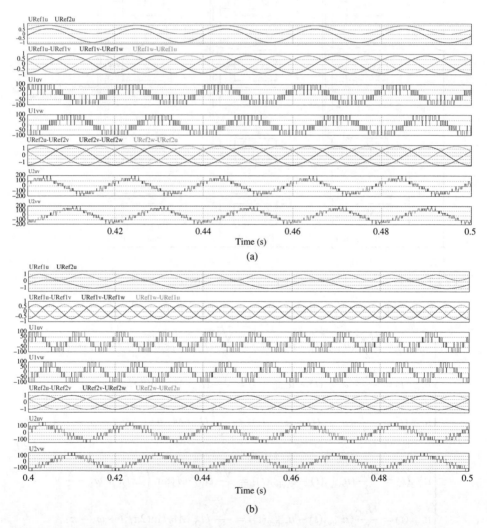

Figure 3.10 Simulation waveforms of the single-input dual-output three-phase DC–AC converter. (a) EF′ mode. (b) DF mode.

When $u_{Reful} = 0.4\sin 200\pi t + 0.6$ and $u_{Refu2} = 0.6\sin 100\pi t - 0.4$, the corresponding waveforms in Figure 3.10b prove that the proposed converter operates in the DF mode.

From the above analysis, the proposed single-input dual-output three-phase DC–AC converter can operate as two modular multilevel converters (MMCs) with three common arms. Compared with two separate MMCs, the proposed converter could save 25% of the semiconductors and 50% of the buffer inductors.

3.5 Single-Input Multiple-Output Half-Bridge Single-Phase DC–AC Converter

3.5.1 Basic Structure and Operating Principle

If the middle switching arm in Figure 3.1 is replaced by several switching arms connected in series, then a single-input multiple-output half-bridge single-phase DC–AC converter can be constructed [5]. The proposed converter with M outputs is shown in Figure 3.11, which has M+1 switching arms and M+1 input capacitors (C_1, C_2, ..., C_{M+1}). Each switching arm consists of N SMs connected in series, and the structure of the SM can be the same as that in Figure 3.2.

Based on Figure 3.11, the kth output voltage can be expressed by

$$u_k = u_{ak} - u_{bk} = u_{ak} - \frac{M+1-k}{M+1}U_{DC} \tag{3.22}$$

where $k = 1, 2, ..., M$.

Assuming that the primary inductance L_p, the secondary inductance L_s, and the mutual inductance M_L of the coupled inductor are identical, that is $L_p = L_s = M_L$, we have

$$u_{Lp} = u_{Ls} \tag{3.23}$$

From Figure 3.11, the relationship between joint point voltages can be defined by

$$\begin{cases} u_{a1} = U_{DC} - u_{A1} - u_{Lp} \\ u_{aM} = u_{A(M+1)} + u_{Ls} \end{cases} \tag{3.24}$$

$$u_{a1} - u_{aM} = u_{A2} + u_{A3} + \cdots + u_{AM} \tag{3.25}$$

Based on Eq. (3.23), we have

$$u_{a1} + u_{aM} = U_{DC} - u_{A1} + u_{A(M+1)} \tag{3.26}$$

Then the voltage on joint point a_1 is

$$u_{a1} = \frac{U_{DC} - u_{A1} + u_{A2} + u_{A3} + \cdots + u_{A(M+1)}}{2} \tag{3.27}$$

and the other joint point voltage can be determined by

$$u_{ax} = u_{a1} - \sum_{i=2}^{x} u_{Ai} \tag{3.28}$$

where $x = 2, 3, ..., M$, u_{Ai} is the output voltage of the ith switching arm.

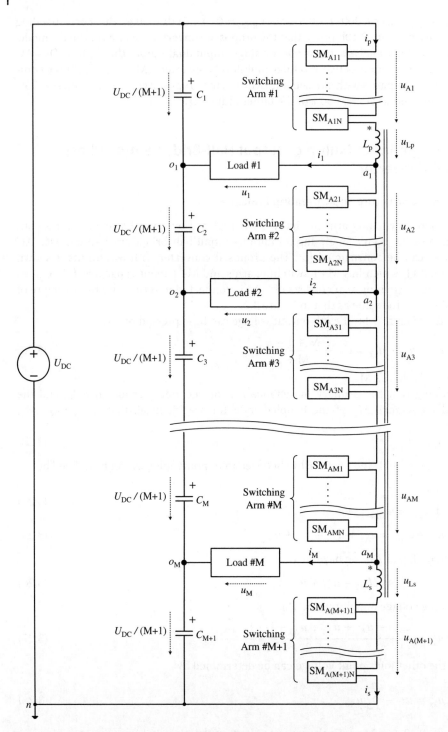

Figure 3.11 Single-input multiple-output half-bridge single-phase DC–AC converter.

Combining Eqs. (3.27) and (3.28), the general form of the joint point voltage in Figure 3.11 is

$$u_{ak} = \frac{1}{2}\left(U_{DC} - \sum_{i=1}^{k} u_{Ai} + \sum_{j=k+1}^{M+1} u_{Aj} \right) \tag{3.29}$$

Substituting Eq. (3.29) into Eq. (3.22), the kth output voltage equals

$$u_k = \frac{U_{DC}}{2}\left[\frac{1}{U_{DC}}\left(\sum_{j=k+1}^{M+1} u_{Aj} - \sum_{i=1}^{k} u_{Ai} \right) - \left(1 - \frac{2k}{M+1} \right) \right] \tag{3.30}$$

Therefore, the desired AC output voltage can be obtained by controlling the switching-arm voltages. Because $u_{k-1} - u_k = u_{Ak} - \frac{U_{DC}}{M+1}$ can be obtained from Eq. (3.30), the switching-arm voltage will satisfy the following equation at steady state:

$$u_{Ak} = u_k - u_{k-1} - \frac{U_{DC}}{M+1} \tag{3.31}$$

3.5.2 Control Scheme

When the CPS-SPWM scheme is applied to the proposed single-input multiple-output half-bridge single-phase DC–AC converter, M modulating references are needed. Based on Eq. (3.30), the kth reference signal can be defined by

$$
\begin{aligned}
u_{Refk}(t) &= \frac{1}{U_{DC}}\left(\sum_{j=k+1}^{M+1} u_{Aj} - \sum_{i=1}^{k} u_{Ai} \right) \\
&= \frac{2U_k}{U_{DC}}\sin(2\pi f_k t + \varphi_k) + \left(1 - \frac{2k}{M+1} \right) \\
&= M_k \sin(2\pi f_k t + \varphi_k) + U_{osk}
\end{aligned} \tag{3.32}
$$

where M_k, f_k, φ_k, and U_{osk} are the modulated ratio, frequency, phase angle, and DC offset of the kth reference signal, respectively.

If there are N SMs in each switching arm, then N carrier signals phase shifted by an angle $360°/N$ are needed. Therefore, the control signals of the ith SM in all switching arms are generated by comparing the reference signals with a common carrier signal u_{Ci}, where $i = 1, 2, \ldots, N$, respectively. That is, the kth reference signal u_{Refk} is compared with the carrier signal in the kth comparator $COMP_k$. If the control signal is used to control switch T_1 in the SM, then according to Eq. (3.31), the control signal for the kth switching-arm voltage u_{gAki} is obtained by the XOR operation of the output of $COMP_{k-1}$ and the inverted output of $COMP_k$, while the control signal for the first switching arm (#1) u_{gA1i} is the output of $COMP_1$ and the control signal for the last switching arm (#M+1); $u_{gA(M+1)i}$ is just the inverted output of $COMP_M$.

The control schematic described above is shown in Figure 3.12. Because the reference signals satisfy $u_{Ref1}(t) > u_{Ref2}(t)$, it is found that only one control signal in Figure 3.12 is at low level or equal to zero. As a result, there are N SMs with output voltage U_C and $M \cdot N$ SMs with zero output voltage at any moment. The capacitor voltage should keep at $U_C = U_{DC}/N$ to ensure the output voltage sum of all switching arms satisfies $\sum_{i=1}^{M+1} u_{Ai} = U_{DC}$.

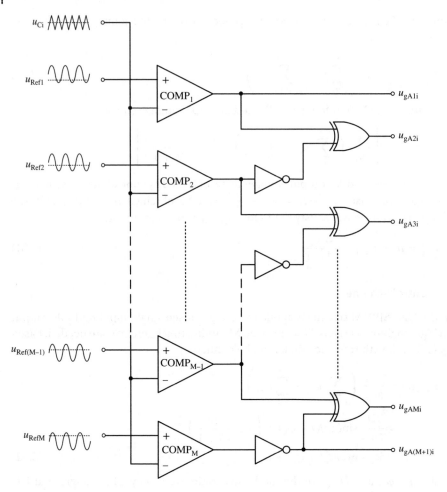

Figure 3.12 Control scheme for the single-input multiple-output half-bridge single-phase DC–AC converter.

3.5.3 Output Voltage Verification

It is obvious that the proposed single-input multiple-output half-bridge single-phase DC–AC converter can operate in both EF and DF modes. A three-output simulation prototype (M = 3) is used as an example to verify the performance of the proposed converter.

Based on Eq. (3.31), the relationship between one carrier signal u_{Ci} and three reference signals under different operating modes is demonstrated in Figure 3.13. The modulated ratio and DC offset in EF′, EF, and DF modes should satisfy $M_k \leq 1 - |U_{osk}|$ and $U_{osk} - M_k \geq U_{os(k+1)} + M_{k+1}$ to avoid over-modulation and overlap of reference signals.

By selecting N = 4, $U_{DC} = 400$ V, $L_{AC} = 0.1$ H, and $R_{AC} = 10\Omega$ to be the simulation parameters, the simulation waveforms are shown in Figure 3.14. When $u_{Ref1} = \frac{1}{2}\sin 100\pi t + \frac{1}{2}$, $u_{Ref2} = \frac{2}{3}\sin 100\pi t$, and $u_{Ref3} = \frac{1}{4}\sin 100\pi t - \frac{1}{2}$, the reference signals $u_{Ref1} \sim u_{Ref3}$, the first output voltage u_1 and the output current i_1, the second

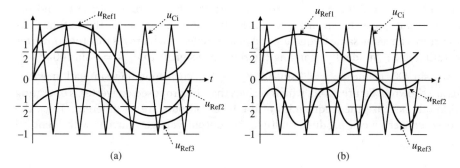

Figure 3.13 Relationship between carrier signal and three reference signals under different operating modes. (a) EF′. (b) DF.

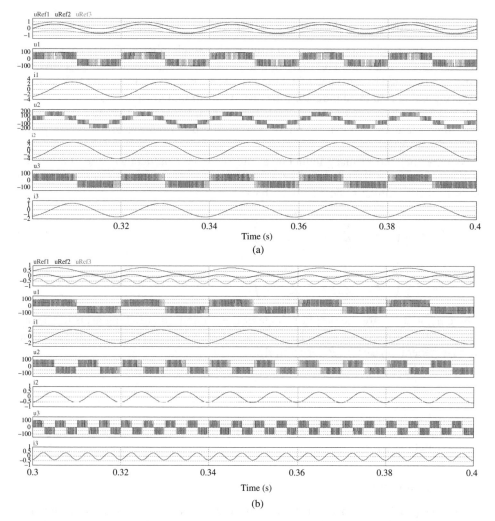

Figure 3.14 Simulation waveforms of the single-input three-output half-bridge single-phase DC–AC converter. (a) EF′ mode. (b) DF mode.

output voltage u_2 and the output current i_2, and the third output voltage u_3 and the output current i_3 are shown in Figure 3.14a. It is shown that all of the output voltages/currents have the same frequency and phase angle, which proves that the converter is operating in EF′ mode. By changing the reference signals to $u_{Ref1} = \frac{1}{3}\sin 100\pi t + \frac{1}{2}$, $u_{Ref2} = \frac{1}{6}\sin 200\pi t$, and $u_{Ref3} = \frac{1}{4}\sin 400\pi t - \frac{1}{2}$, the corresponding simulation waveforms are shown in Figure 3.14b, which verifies that the proposed converter is operating in DF mode because the frequencies of the three outputs are all different. Thus, the proposed converter can produce multiple AC output voltages/current with identical or DF.

3.6 Single-Input Multiple-Output Full-Bridge Single-Phase DC–AC Converter

3.6.1 Basic Structure and Operating Principle

Similar to the single-input dual-output full-bridge single-phase DC–AC converter in Section 3.3, the single-input multiple-output full-bridge single-phase DC–AC converter can be constructed by replacing M + 1 input capacitors in Figure 3.12 with M + 1 switching arms and one coupled inductor [6], which is illustrated in Figure 3.15.

If M + 1 switching arms and one coupled inductor are considered as one general phase unit, then there are two general phase units in the proposed full-bridge single-phase DC–AC converter.

Assume that the primary inductance, the secondary inductance, and the mutual inductance of the coupled inductor are the same, that is, $L_{pa} = L_{sa} = M_a$ and $L_{pb} = L_{sb} = M_b$. Similar to Eq. (3.29), we have

$$u_{ak} = \frac{1}{2}\left(U_{DC} - \sum_{i=1}^{k} u_{Ai} + \sum_{j=k+1}^{M+1} u_{Aj} \right) \tag{3.33}$$

$$u_{bk} = \frac{1}{2}\left(U_{DC} - \sum_{i=1}^{k} u_{Bi} + \sum_{j=k+1}^{M+1} u_{Bj} \right) \tag{3.34}$$

where u_{Ai} and u_{Aj} are the output voltage of the ith and jth switching arms in general phase unit A, respectively. u_{Bi} and u_{Bj} are the output voltage of the ith and jth switching arms in general phase unit B, respectively.

Based on Figure 3.15, the kth output voltage can be expressed by

$$u_k = u_{ak} - u_{bk} \tag{3.35}$$

where $k = 1, 2, \ldots, M$.

Thus, the desired AC output voltage can be obtained by controlling the output voltages of the switching arms.

3.6.2 Control Scheme

Based on Eq. (3.33), if both u_{ak} and u_{bk} are sinusoidal signals with the same frequency and DC offset, then the output voltage u_k will be a sinusoidal signal. In order to simplify the

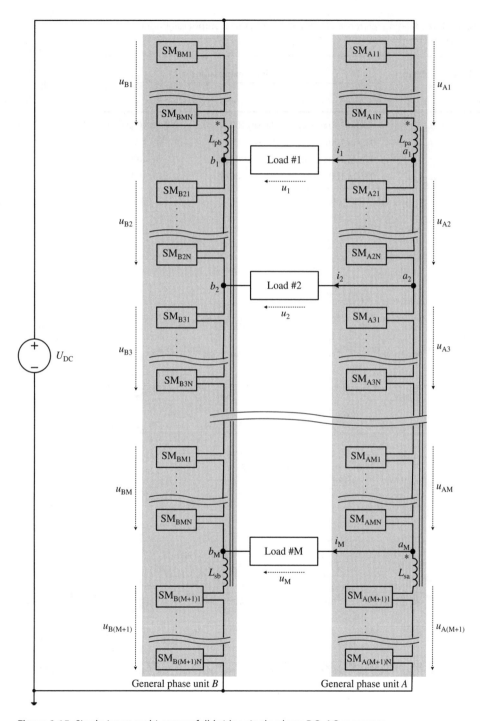

Figure 3.15 Single-input multi-output full-bridge single-phase DC–AC converter.

Table 3.1 Modulating reference of the single-input three-output full-bridge single-phase DC–AC converter.

Mode	Reference	Voltage expressions		
		$k = 1$	$k = 2$	$k = 3$
EF′	u_{Refak}	$0.3 \sin 100\pi t + 0.7$	$0.6 \sin 100\pi t + 0.3$	$0.9 \sin 100\pi t - 0.7$
	u_{Refk}	$0.6 \sin 100\pi t$	$1.2 \sin 100\pi t$	$1.8 \sin 100\pi t$
DF	u_{Refak}	$0.3 \sin 100\pi t + 0.7$	$0.4 \sin 200\pi t$	$0.2 \sin 400\pi t - 0.8$
	u_{Refk}	$0.6 \sin 100\pi t$	$0.8 \sin 200\pi t$	$0.4 \sin 400\pi t$

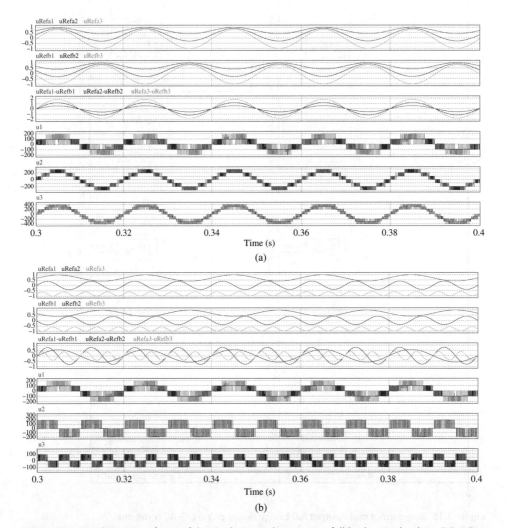

Figure 3.16 Simulation waveforms of the single-input three-output full-bridge single-phase DC–AC converter. (a) EF′ mode. (b) DF mode.

control scheme and improve the magnitude of the fundamental waveform, the reference signals for u_{ak} and u_{bk} can be defined by

$$u_{Refak}(t) = M_k \sin(2\pi f_k t + \varphi_k) + U_{osk} \tag{3.36}$$

$$u_{Refbk}(t) = -M_k \sin(2\pi f_k t + \varphi_k) + U_{osk} \tag{3.37}$$

where M_k is modulated ratio, f_k is frequency, φ_k is phase angle, U_{osk} is DC offset, and $M_k + U_{osk} \leq 1$.

Thus, the reference signal according to the kth output voltage is expressed by

$$u_{Refk}(t) = u_{Refak}(t) - u_{Refbk}(t) = 2M_k \sin(2\pi f_k t + \varphi_k) \tag{3.38}$$

By applying the CPS-SPWM scheme to the proposed single-input multiple-output full-bridge single-phase DC–AC converter, the control signals for the SMs in each phase unit can be generated by the same method as in Figure 3.12. Normally, $u_{Refak} \geq u_{Refa(k+1)}$ are selected to avoid crossover intersection of reference voltages.

3.6.3 Output Voltage Verification

A three-output simulation prototype (M = 3) is used as an example to verify the performance of the proposed single-input multiple-output full-bridge single-phase DC–AC converter. The prototype parameters include N = 4, $U_{DC} = 400$ V, $L_{AC} = 0.1$ H, and $R_{AC} = 10\Omega$; the definitions of the reference signals are listed in Table 3.1.

The corresponding simulation waveforms of the reference signals $u_{Refa1} \sim u_{Refa3}$ for general phase unit A, the reference signals $u_{Refb1} \sim u_{Refb3}$ for general phase unit B, the first output voltage u_1, the second output voltage u_2, and the third output voltage u_3 are shown in Figure 3.16.

Figure 3.16a shows that the proposed converter is operating in EF′ mode, because the frequencies and phase shifts of the three outputs are the same. However, Figure 3.16b shows that the proposed converter is operating in DF mode, because the frequencies of the three outputs are all different.

3.7 Single-Input Multiple-Output Three-Phase DC–AC Converter

3.7.1 Basic Structure and Operating Principle

The single-input multiple-output three-phase DC–AC converter can be constructed using three general switching units, as shown in Figure 3.17 [7]. The structure of the three general phase units (U, V, W) is the same, made up of M + 1 switching arms and one coupled inductor. Similar to the converter proposed in Section 3.5, each switching arm is made up of N SMs connected in series, and the SM shown in Figure 3.2 is selected.

Similar to Section 3.6, the phase voltage of the kth output can be defined by

$$u_{xk} = \frac{1}{2}\left(U_{DC} - \sum_{i=1}^{k} u_{Xi} + \sum_{j=k+1}^{M+1} u_{Xj}\right) \tag{3.39}$$

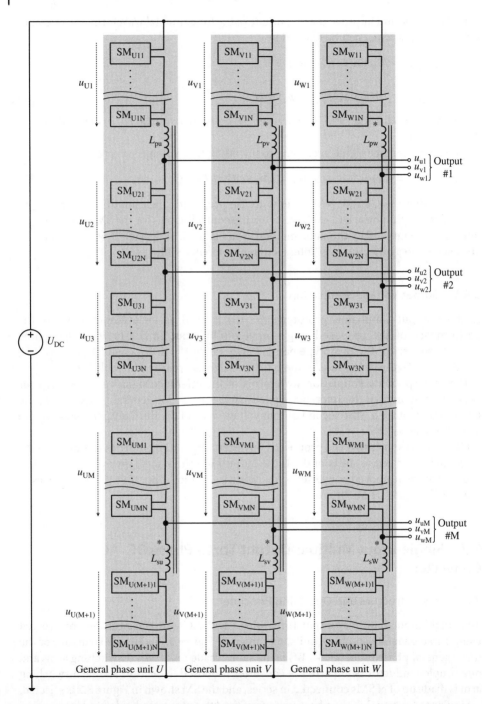

Figure 3.17 Single-input multi-output three-phase DC–AC converter.

where $x = u$, v, w, $k = 1$, 2, …, M, u_{Xi} and u_{Xj} are the output voltage of the ith and jth switching arms in general phase unit X (X $= U$, V, W), respectively.

Thus, the line voltage of the kth output is

$$\begin{cases} u_{uvk} = u_{uk} - u_{vk} \\ u_{vwk} = u_{vk} - u_{wk} \\ u_{wuk} = u_{wk} - u_{uk} \end{cases} \tag{3.40}$$

3.7.2 Control Scheme

Similar to Section 3.4.2, the reference signal can be defined in the following equation if a three-phase sinusoidal output is desired:

$$u_{Refxk}(t) = M_k \sin(2\pi f_k t + \varphi_k + \phi_x) + U_{osk} \tag{3.41}$$

where $\phi_u = 0$, $\phi_v = -\frac{2\pi}{3}$, $\phi_w = \frac{2\pi}{3}$, M_k, f_k, φ_k, and U_{osk} are the modulated ratio, frequency, phase shift, and DC offset of the kth reference signal, respectively. And the relationship between M_k and U_{osk} should agree with $M_k + U_{osk} \leq 1$.

Applying the CPS-SPWM scheme to the proposed single-input multiple-output three-phase DC–AC converter, the control signals for the SMs can be generated by the same method as in Figure 3.12. The proposed converter can operate in EF′, EF, and DF modes as well, while $u_{Refxk} \geq u_{Refx(k+1)}$ are selected to avoid crossover intersection of reference voltages.

3.7.3 Output Voltage Verification

A three-output simulation prototype (M $= 3$) is used as an example to verify the performance of the proposed single-input multiple-output full-bridge single-phase DC–AC converter. N $= 4$, $U_{DC} = 400$ V, $f_C = 900$ Hz, $L_{AC} = 0.1$ H and $R_{AC} = 10\Omega$ are selected as simulation parameters. The definitions of reference signals are listed in Table 3.2.

Table 3.2 Modulating reference of the proposed single-input three-output three-phase DC–AC converter.

Mode	Output phase x	Expressions of modulating reference u_{Refxk}		
		$k = 1$	$k = 2$	$k = 3$
EF′	u	$0.6\sin 100\pi t + 0.4$	$0.8\sin 100\pi t$	$0.4\sin 100\pi t - 0.6$
	v	$0.6\sin\left(100\pi t - \frac{2\pi}{3}\right) + 0.4$	$0.8\sin\left(100\pi t - \frac{2\pi}{3}\right)$	$0.4\sin\left(100\pi t - \frac{2\pi}{3}\right) - 0.6$
	w	$0.6\sin\left(100\pi t + \frac{2\pi}{3}\right) + 0.4$	$0.8\sin\left(100\pi t + \frac{2\pi}{3}\right)$	$0.4\sin\left(100\pi t + \frac{2\pi}{3}\right) - 0.6$
DF	u	$0.3\sin 100\pi t + 0.7$	$0.5\sin 200\pi t - 0.1$	$0.2\sin 300\pi t - 0.8$
	v	$0.3\sin\left(100\pi t - \frac{2\pi}{3}\right) + 0.7$	$0.5\sin\left(200\pi t - \frac{2\pi}{3}\right) - 0.1$	$0.2\sin\left(300\pi t - \frac{2\pi}{3}\right) - 0.8$
	w	$0.3\sin\left(100\pi t + \frac{2\pi}{3}\right) + 0.7$	$0.5\sin\left(200\pi t + \frac{2\pi}{3}\right) - 0.1$	$0.2\sin\left(300\pi t + \frac{2\pi}{3}\right) - 0.8$

The corresponding simulation waveforms of the reference signals $u_{\text{Refu1}} \sim u_{\text{Refu3}}$ for phase unit U, three output reference signals $(u_{\text{Refu1}} - u_{\text{Refv1}})$, $(u_{\text{Refu2}} - u_{\text{Refv2}})$, and $(u_{\text{Refu3}} - u_{\text{Refv3}})$, the first output line voltage u_{uv1}, the second output line voltage u_{uv2}, and the third output line voltage u_{uv3} are shown in Figure 3.18. Figure 3.18a shows that the proposed converter is operating in EF' mode, because the frequencies of the three outputs are the same. However, Figure 3.18b shows that the proposed converter is operating in DF mode, because the frequencies of the three outputs are all different.

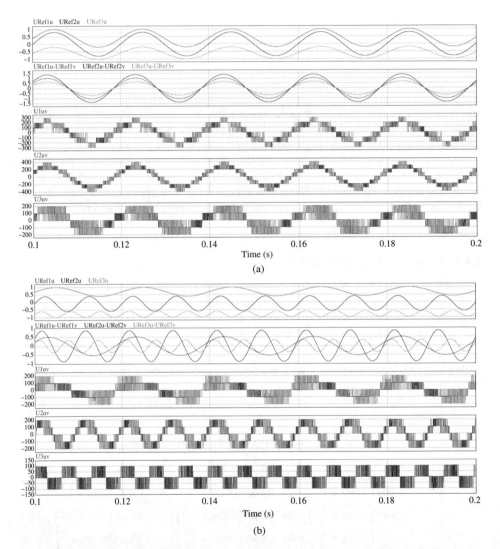

Figure 3.18 Simulation waveforms of the proposed single-input three-output three-phase DC–AC converter. (a) EF' mode. (b) DF mode.

3.8 Summary

Six kinds of single-input multiple-output high-voltage DC–AC inverter have been proposed in this chapter, with the following characteristics:

1) Two or more single-phase/three-phase AC outputs can be obtained.
2) Output frequencies can be identical or different.
3) The switching arm is made up of N SMs, and the output voltage of the converter has multiple levels and low harmonics.
4) The voltage stress of each power switch in the SM is only U_{DC}/N, thus the converter can operate in high-voltage situations.

Compared with the traditional multilevel inverter and MMC, the proposed single-input multiple-output high-voltage DC–AC inverter simplifies the topology, minimizes the number of switching components, and reduces the cost, which is suitable for high-voltage high-power multiple-output applications.

References

1 Zhang, B., Fu, J., Qiu, D. Y. Double-output single-phase three switching-groups MMC inverter without DC bias and its control method. State Intellectual Property Office of the P.R.C., ZL 201410042990.5, 2016.4.13.

2 Zhang, B., Fu, J., Qiu, D. Y. Double-output single-phase six switching-groups MMC inverter and its control method. State Intellectual Property Office of the P.R.C., ZL 201410042070.3, 2016.6.22.

3 Zhang, B., Fu, J., Qiu, D. Y. Double-load three-phase nine switching-groups MMC inverter and its control method. State Intellectual Property Office of the P.R.C., ZL 201410042954.9, 2017.1.11.

4 Fu, J., Zhang, B., Qiu, D. Y. (2014) A novel nine-arm modular multilevel converter. Proceedings of the 40th Annual Conference of the IEEE Industrial Electronics Society (IECON), Dallas, TX, October 29–November 1, 4528–4533.

5 Zhang, B., Fu, J., Qiu, D. Y. N-output single-phase N+1 switching-groups MMC inverter and its control method. State Intellectual Property Office of the P.R.C., ZL 201410042873.9, 2016.6.22.

6 Zhang, B., Fu, J., Qiu, D. Y. N-output single-phase 2N+2 switching-groups MMC inverter and its control method. State Intellectual Property Office of the P.R.C., ZL 201410042826.4, 2017.4.12.

7 Zhang, B., Fu, J., Qiu, D. Y. N-output three-phase 3N+3 switching-groups MMC inverter and its control method. State Intellectual Property Office of thc P.R.C., ZL 201410042730.8, 2017.1.18.

4

Multiple-Input Single-Output High-voltage AC–DC Converters

4.1 Introduction

In the high-voltage direct current (HVDC) transmission system, there are several AC input sources which need to be converted to DC form; the common solution is to apply one AC–DC converter for each AC source and then connect all of the DC output sides together. In order to simplify the structure of the whole system, a series of AC–DC converters with multiple AC inputs and single DC output are proposed in this chapter.

Based on the symmetry between inverter (DC–AC converter) and rectifier (AC–DC converter), the multiple-input single-output AC–DC converter can be obtained just by exchanging the input and output sides of the single-input multiple-output DC–AC converter presented in Chapter 3. Thus, it is the AC voltage sources which are connected between the switching arms, while the DC output load is in parallel with the switching arms. The circuit topology and operating principle are described in detail, the carrier phase-shifted sinusoidal pulse-width modulation (CPS-SPWM) scheme is applied to the proposed converters, and the simulation waveforms are provided to verify the feasibility of the proposed multiple-input single-output AC–DC converters.

4.2 Single-Phase Three-Arm Dual-Input Single-Output AC–DC Converter

4.2.1 Basic Structure and Operating Principle

Similar to the single-input dual-output half-bridge single-phase DC–AC converter in Figure 3.1, the proposed single-phase three-arm dual-input single-output AC–DC converter is shown in Figure 4.1, composed of three switching arms (U, M, and L), three DC capacitors (C_1, C_2, and C_3), and one coupled inductor as well [1]. The structures of switching arm and sub-module (SM) are the same as those presented in Chapter 3. Different from Figure 3.1, the input AC source in series with an inductor is connected between the switching arm and the capacitor, and the output load is in parallel with the capacitor string.

From Figure 4.1, the voltages at two terminal points of the middle switching arm (M) can be expressed by

$$\begin{cases} u_{a1} = U_O - u_U - u_{Lp} \\ u_{a2} = u_L + u_{Ls} \end{cases} \tag{4.1}$$

Multi-terminal High-voltage Converter, First Edition. Bo Zhang and Dongyuan Qiu.
© 2019 John Wiley & Sons Singapore Pte. Ltd. Published 2019 by John Wiley & Sons Singapore Pte. Ltd.

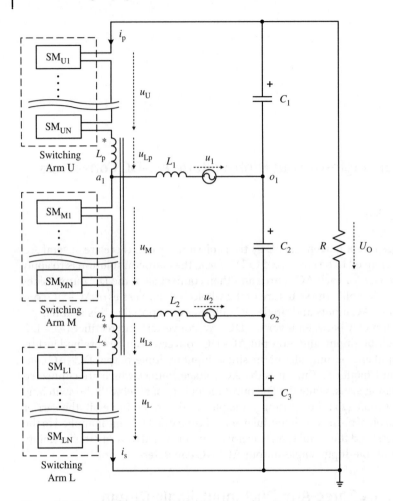

Figure 4.1 Single-phase three-arm dual-input single-output AC–DC converter.

and their voltage difference is determined by

$$u_{a1} - u_{a2} = u_M \qquad (4.2)$$

where U_O is the DC output voltage, u_{Lp} and u_{Ls} are the voltages on the primary and secondary windings of the coupled inductor, u_U, u_M, and u_L are the output voltages of the upper switching arm (U), the middle switching arm (M), and the lower switching arm (L), respectively.

Assuming that the primary inductance L_p, the secondary inductance L_s, and the mutual inductance M_L of the coupled inductor are equal, that is, $L_p = L_s = M_L$, we have

$$u_{Ls} = u_{Lp} \qquad (4.3)$$

Based on Eqs. (4.2) and (4.3), Eq. (4.1) turns out to be

$$\begin{cases} u_{a1} = \dfrac{U_O + u_L + u_M - u_U}{2} \\[2mm] u_{a2} = \dfrac{U_O + u_L - u_M - u_U}{2} \end{cases} \qquad (4.4)$$

According to Kirchhoff's voltage law, the following voltage equations at steady state are established:

$$\begin{cases} u_{a1o1} = u_{a1} - u_{o1} = u_{a1} - \dfrac{2}{3}U_O \\ u_{a2o2} = u_{a2} - u_{o2} = u_{a2} - \dfrac{1}{3}U_O \end{cases}$$

(4.5)

where u_{a1o1} and u_{a2o2} are the voltages between switching arms and capacitors.

By substituting Eq. (4.4) into Eq. (4.5), the DC output voltage can be defined by

$$U_O = 3(u_L + u_M - u_U) - 6u_{a1o1} = 6u_{a2o2} - 3(u_L - u_M - u_U)$$

(4.6)

4.2.2 Control Scheme

In order to obtain a constant output DC voltage U_O, based on Eq. (4.6), the reference signals for controlling the output voltages of three switching arms (u_U, u_M, and u_L) should satisfy the following equations:

$$\begin{cases} u_{Ref1} = \dfrac{u_L + u_M - u_U}{U_O} = \dfrac{2u_{a1o1}}{U_O} + \dfrac{1}{3} \\ u_{Ref2} = \dfrac{u_L - u_M - u_U}{U_O} = \dfrac{2u_{a2o2}}{U_O} - \dfrac{1}{3} \end{cases}$$

(4.7)

As the function of the series inductors L_1 and L_2 is to keep the voltage balance and the average value of the inductor voltage in steady state equal to zero, Eq. (4.7) turns out to be

$$\begin{cases} u_{Ref1} = \dfrac{2u_1}{U_O} + \dfrac{1}{3} = M_1 \sin(2\pi f_1 t + \varphi_1) + \dfrac{1}{3} \\ u_{Ref2} = \dfrac{2u_2}{U_O} - \dfrac{1}{3} = M_2 \sin(2\pi f_2 t + \varphi_2) - \dfrac{1}{3} \end{cases}$$

(4.8)

where $M_1 = \dfrac{2U_1}{U_O}$ and $M_2 = \dfrac{2U_2}{U_O}$ are modulated ratios, while U_1 and U_2 are amplitudes, f_1 and f_2 are frequencies, φ_1 and φ_2 are phase shifts of the dual input voltages, respectively.

It is known that there are two operating modes according to the frequency of the input voltage. One is the different frequency (DF) mode and the other is the equal frequency (EF) mode. However, there is a special case in the EF mode in which both of the input voltages are in phase, and it is named the "EF′ mode" in this chapter.

If the CPS-SPWM scheme is selected to control the proposed AC–DC converter, similar to Figure 3.3, the reference signals should meet the requirements of $1 \ge u_{Ref1} \ge u_{Ref2} \ge -1$ in order to avoid their overlap. Thus, the relationship of two modulating ratios is described by

$$\begin{cases} M_1 + M_2 \le \dfrac{2}{3}, & \text{for DF and EF modes, except for EF′ mode} \\ M_1, M_2 \le \dfrac{2}{3}, & \text{only for EF′ mode} \end{cases}$$

(4.9)

Table 4.1 Parameters of the single-phase three-arm dual-input single-output AC–DC converter.

Mode	U_O (V)	U_1 (V)	f_1 (Hz)	φ_1 (rad)	M_1	U_2 (V)	f_2 (Hz)	φ_2 (rad)	M_2
DF	300	37.5	50	0	$\frac{1}{4}$	30	100	0	$\frac{1}{5}$
EF′	300	75	50	0	$\frac{1}{2}$	100	50	0	$\frac{2}{3}$

and the output voltage value will be in the range of

$$\begin{cases} U_O \geq 3(U_1 + U_2), \text{for DF and EF modes, except for EF}' \text{ mode} \\ U_O \geq 3 \times \max(U_1, U_2), \text{only for EF}' \text{ mode} \end{cases} \tag{4.10}$$

4.2.3 Performance Verification

In this section, the feasibility of the proposed single-phase three-arm dual-input single-output AC–DC converter will be proven by simulation results. The control scheme in Figure 3.4 can be used to generate the gate signals for SMs in switching arms, and the simulation parameters in different modes are listed in Table 4.1.

When the number of SMs in each switching arm is $N=4$, the simulation waveforms in DF mode are shown in Figure 4.2. From Figure 4.2a, it is found that the reference signals u_{Ref1} and u_{Ref2} defined by Eq. (4.8) do not overlap and the output voltage remains at about $300\,\text{V}$, which agrees with $U_O = \frac{2U_1}{M_1} = \frac{2U_2}{M_2} = 300\,\text{V}$. The waveforms of the switching-arm voltage $u_L - u_M - u_U$ and the first input voltage u_1, $u_L + u_M - u_U$ and the second input voltage u_2 are shown in Figure 4.2b, while their spectrums are shown in Figure 4.2c. It is obvious that $u_L + u_M - u_U$ has a DC component with value $100\,\text{V}$ (or $\frac{U_O}{3}$) and a fundamental component twice the input voltage u_1, while $u_L - u_M - u_U$ has a DC component the same as that of $u_L + u_M - u_U$ but a fundamental component twice the second input voltage u_2. Therefore, the simulation results are consistent with Eq. (4.6).

Similarly, the corresponding waveforms in EF′ mode are shown in Figure 4.3, in which $U_O = 3U_1 = 300\,\text{V}$ is in agreement with Eq. (4.10). It is clear that the proposed converter can convert dual AC input voltages with identical or different frequencies to constant DC output voltage.

4.3 Single-Phase Six-Arm Dual-Input Single-Output AC–DC Converter

4.3.1 Basic Structure and Operating Principle

As capacitors with high voltage stress should be used in the single-phase three-arm dual-input single-output AC–DC converter, the application of the proposed three-arm AC–DC converter is not available for the situation in which the output DC voltage U_O is relatively high. By replacing the series capacitors in Figure 4.1 with switching

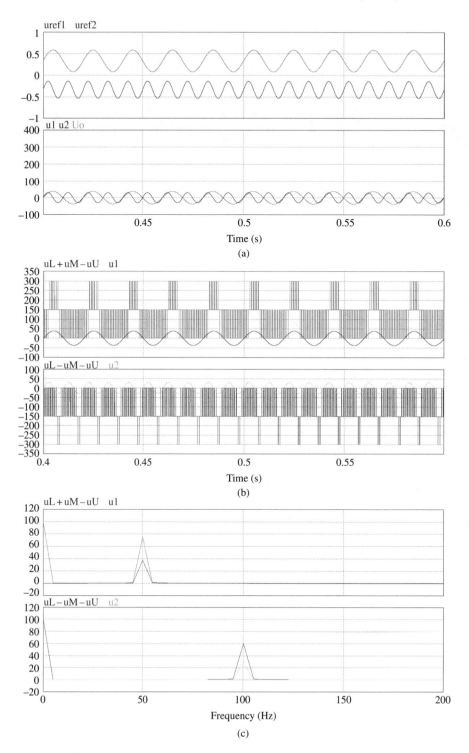

Figure 4.2 Simulation waveforms of the proposed single-phase three-arm dual-input single-output AC–DC converter in DF mode. (a) Waveforms of reference signals, input, and output voltages. (b) Waveforms of $u_L + u_M - u_U$ and u_1, $u_L - u_M - u_U$ and u_2. (c) Spectrums of $u_L + u_M - u_U$ and u_1, $u_L - u_M - u_U$ and u_2.

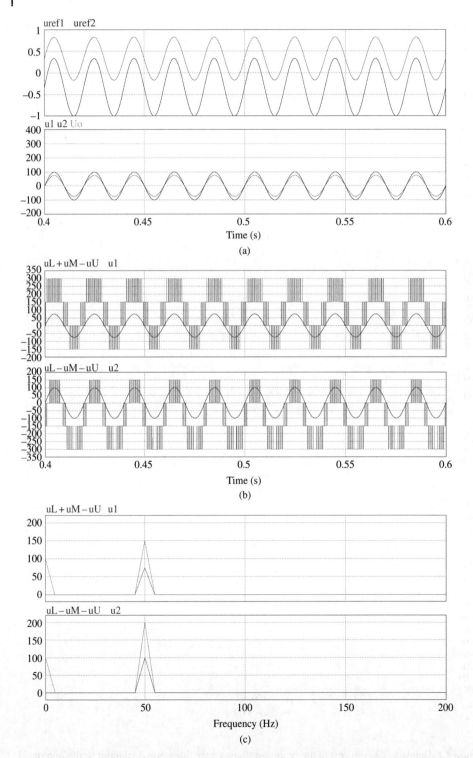

Figure 4.3 Simulation waveforms of the proposed single-phase three-arm dual-input single-output AC–DC converter in EF′ mode. (a) Waveforms of reference signals, input, and output voltages. (b) Waveforms of $u_L + u_M - u_U$ and u_1, $u_L - u_M - u_U$ and u_2. (c) Spectrums of $u_L + u_M - u_U$ and u_1, $u_L - u_M - u_U$ and u_2.

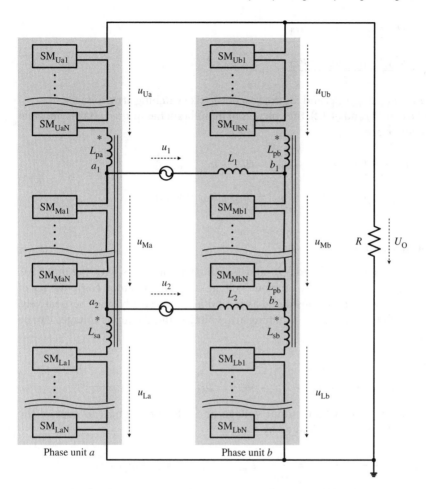

Figure 4.4 Single-phase six-arm dual-input single-output AC–DC converter.

arms, a single-phase six-arm dual-input single-output AC–DC converter is obtained, as shown in Figure 4.4 [2].

By defining three switching arms and one coupled inductor which are connected in series as one phase unit, the proposed six-arm single-phase AC–DC converter consists of two phase units a and b, and the input voltage sources are connected between the phase units.

Similar to the analysis for the single-phase three-arm dual-input single-output AC–DC converter in Section 4.2.1, the terminal point voltages of the middle switching arm can be expressed by

$$\begin{cases} u_{a1} = \dfrac{U_O + u_{La} + u_{Ma} - u_{Ua}}{2} \\[2mm] u_{a2} = \dfrac{U_O + u_{La} - u_{Ma} - u_{Ua}}{2} \end{cases} \qquad (4.11)$$

$$\begin{cases} u_{b1} = \dfrac{U_O + u_{Lb} + u_{Mb} - u_{Ub}}{2} \\[2mm] u_{b2} = \dfrac{U_O + u_{Lb} - u_{Mb} - u_{Ub}}{2} \end{cases} \tag{4.12}$$

where u_{Ua}, u_{Ma}, u_{La}, u_{Ub}, u_{Mb}, and u_{Lb} represent the six switching arm voltages.

Based on Eqs. (4.11) and (4.12), the voltages of input branches or the voltages between phase unit a and b are

$$\begin{cases} u_{a1} - u_{b1} = \dfrac{1}{2}[(u_{La} + u_{Ma} - u_{Ua}) - (u_{Lb} + u_{Mb} - u_{Ub})] \\[2mm] u_{a2} - u_{b2} = \dfrac{1}{2}[(u_{La} - u_{Ma} - u_{Ua}) - (u_{Lb} - u_{Mb} - u_{Ub})] \end{cases} \tag{4.13}$$

4.3.2 Control Scheme

As the average voltages of inductors L_1 and L_2, which are in series with the input AC source, equal zero in the period of input voltage, the fundamental components of $u_{a1} - u_{b1}$ and $u_{a2} - u_{b2}$ can be considered as u_1 and u_2 in the steady state, respectively. And the relationship between the switching arm voltages and the input voltages can be expressed by

$$\begin{cases} (u_{La} + u_{Ma} - u_{Ua}) - (u_{Lb} + u_{Mb} - u_{Ub}) = 2u_1 \\[2mm] (u_{La} - u_{Ma} - u_{Ua}) - (u_{Lb} - u_{Mb} - u_{Ub}) = 2u_2 \end{cases} \tag{4.14}$$

In order to obtain a constant output voltage, the reference signals to control the switching arm voltages can be defined by

$$\begin{cases} u_{Refa1} = \dfrac{u_{La} + u_{Ma} - u_{Ua}}{U_O} = \dfrac{u_1}{U_O} + U_{os1} = M_1 \sin(2\pi f_1 t + \varphi_1) + U_{os1} \\[3mm] u_{Refb1} = \dfrac{u_{Lb} + u_{Mb} - u_{Ub}}{U_O} = -\dfrac{u_1}{U_O} + U_{os1} = -M_1 \sin(2\pi f_1 t + \varphi_1) + U_{os1} \end{cases} \tag{4.15}$$

$$\begin{cases} u_{Refa2} = \dfrac{u_{La} - u_{Ma} - u_{Ua}}{U_O} = \dfrac{u_2}{U_O} + U_{os2} = M_2 \sin(2\pi f_2 t + \varphi_2) + U_{os2} \\[3mm] u_{Refb2} = \dfrac{u_{Lb} - u_{Mb} - u_{Ub}}{U_O} = -\dfrac{u_2}{U_O} + U_{os2} = -M_2 \sin(2\pi f_2 t + \varphi_2) + U_{os2} \end{cases} \tag{4.16}$$

where $M_1 = \frac{U_1}{U_O}$ and $M_2 = \frac{U_2}{U_O}$ are modulated ratios, while U_1 and U_2 are amplitudes, f_1 and f_2 are frequencies, φ_1 and φ_2 are phase shifts of the dual input AC voltage sources, respectively. U_{os1} and U_{os2} are DC offsets and satisfy the following equations to avoid over-modulation:

$$\begin{cases} 0 < U_{os1} \leq 1 - M_1 \\[2mm] 0 > U_{os2} \geq M_2 - 1 \end{cases} \tag{4.17}$$

If the CPS-SPWM scheme is selected to control the proposed AC–DC converter, similar to Figure 3.7, then the reference signals should meet the requirements of $u_{\text{Refa1}} \geq u_{\text{Refa2}}$ and $u_{\text{Refb1}} \geq u_{\text{Refb2}}$ in order to avoid their overlap.

When the dual input voltages belong to the DF or EF mode (except for the EF′ mode), the relationship of two modulating ratios will be

$$M_1 + M_2 \leq U_{\text{os1}} - U_{\text{os2}} \tag{4.18}$$

Based on Eqs. (4.17) and (4.18), we have $M_1 + M_2 \leq 1$, then the output voltage value will be in the range of

$$U_{\text{O}} \geq U_1 + U_2 \tag{4.19}$$

When the dual input voltages have the same frequency and phase shift, which is the EF′ mode, the modulated ratio and DC offsets of the reference signals should satisfy

$$\begin{cases} M_1 + U_{\text{os1}} \geq M_2 + U_{\text{os2}} \\ -M_1 + U_{\text{os1}} \geq -M_2 + U_{\text{os2}} \end{cases} \Rightarrow |M_1 - M_2| \leq U_{\text{os1}} - U_{\text{os2}} \tag{4.20}$$

and the output voltage is required as

$$U_{\text{O}} \geq \max(U_1, U_2) \tag{4.21}$$

4.3.3 Performance Verification

In this section, the feasibility of the proposed single-phase six-arm dual-input single-output AC–DC converter will be proven by simulation results, where N = 4 and the capacitor voltage of each SM is assumed to be stable. The simulation parameters when the proposed converter operates in the EF′ mode are listed in Table 4.2. The simulation waveforms and spectrums of reference signals u_{Refa1} and u_{Refa2}, input voltages u_1 and u_2, output voltage U_{O}, and switching-arm voltages $u_{\text{La}} + u_{\text{Ma}} - u_{\text{Ua}}$ and $u_{\text{La}} - u_{\text{Ma}} - u_{\text{Ua}}$ are shown in Figure 4.5.

From the simulation results of group #1, it is found that the reference signals u_{Ref1} and u_{Ref2} do not overlap and the output voltage remains at about 200 V (or $U_{\text{O}} = 200$ V) in Figure 4.5a. The spectrum of the switching-arm voltage $u_{\text{La}} + u_{\text{Ma}} - u_{\text{Ua}}$ has a DC component with value 100 V (or $U_{\text{os1}} \cdot U_{\text{O}}$) and a fundamental component the same as that of the input voltage u_1, while the spectrum of $u_{\text{La}} - u_{\text{Ma}} - u_{\text{Ua}}$ has a DC component of 100 V and a fundamental component the same as that of the input voltage u_2.

By applying the parameter of group #2, it is found that the waveforms of $u_{\text{La}} + u_{\text{Ma}} - u_{\text{Ua}}$ and $u_{\text{La}} - u_{\text{Ma}} - u_{\text{Ua}}$ in Figure 4.5c are different from those in

Table 4.2 Parameters of the single-phase six-arm dual-input single-output AC–DC converter in the EF′ mode.

Group	U_{O} (V)	U_1 (V)	M_1	f_1 (Hz)	ϕ_1 (rad)	U_{os1} (V)	U_2 (V)	M_2	f_2 (Hz)	φ_2 (rad)	U_{os2} (V)
1	200	80	0.4	50	0	0.5	50	0.25	50	0	−0.5
2	200	80	0.4	50	0	0.1	50	0.25	50	0	−0.1
3	200	160	0.8	50	0	0.1	100	0.5	50	0	−0.2

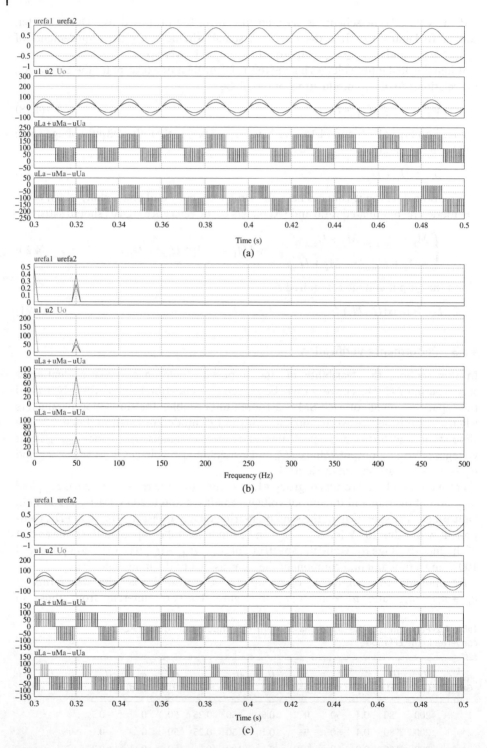

Figure 4.5 Simulation waveforms of the proposed single-phase six-arm dual-input single-output AC–DC converter in the EF' mode. (a) Waveforms of group #1. (b) Spectrums of group #1. (c) Waveforms of group #2. (d) Spectrums of group #2. (e) Waveforms of group #3. (f) Spectrums of group #3.

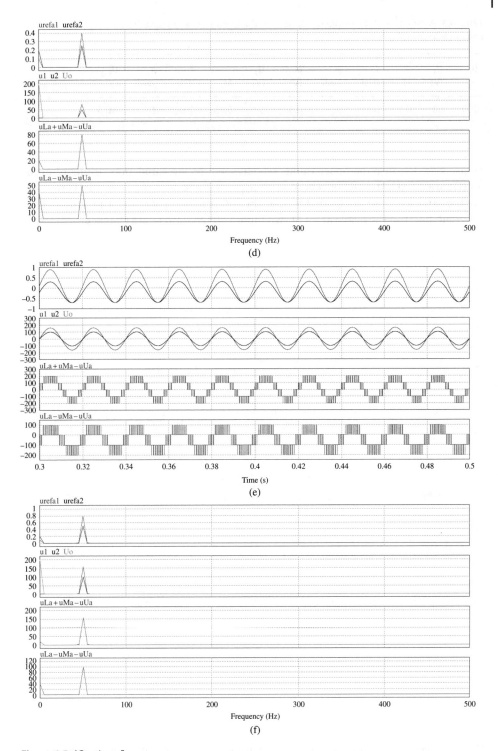

Figure 4.5 (*Continued*)

Table 4.3 Parameters of the single-phase six-arm dual-input single-output AC–DC converter in the DF mode.

U_o (V)	U_1 (V)	M_1	f_1 (Hz)	φ_1 (rad)	U_{os1} (V)	U_2 (V)	M_2	f_2 (Hz)	φ_2 (rad)	U_{os2} (V)
200	80	0.4	50	0	0.5	50	0.25	100	0	−0.5

Figure 4.5a, because the offset values are changed. In the simulation results of group #3 (Figure 4.5e,f), the DC offsets of $u_{La} + u_{Ma} - u_{Ua}$ and $u_{La} - u_{Ma} - u_{Ua}$ in Figure 4.5f are the same as those in Figure 4.5c, but the fundamental component magnitudes are different, because the peak values of two input voltage sources are changed.

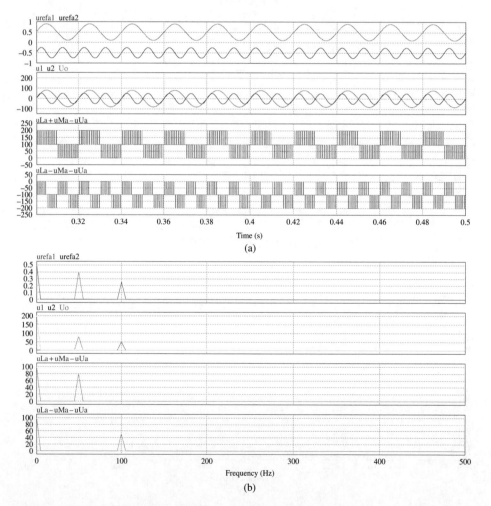

Figure 4.6 Simulation waveforms of the proposed single-phase six-arm dual-input single-output AC–DC converter in the DF mode. (a) Waveforms. (b) Spectrums.

The simulation parameters of the DF mode are listed in Table 4.3, while the simulation waveforms and spectrums of the reference signals u_{Refa1} and u_{Refa2}, input voltages u_1 and u_2, output voltage U_O, and switching-arm voltages $u_{\text{La}} + u_{\text{Ma}} - u_{\text{Ua}}$ and $u_{\text{La}} - u_{\text{Ma}} - u_{\text{Ua}}$ are shown in Figure 4.6.

By selecting the proper DC offsets, both of the reference signals will not be overlapped, and the output voltage will remain at the desired value (or $U_\text{O} = 200$ V). Moreover, the spectrums of $u_{\text{La}} + u_{\text{Ma}} - u_{\text{Ua}}$ and $u_{\text{La}} - u_{\text{Ma}} - u_{\text{Ua}}$ verify Eqs. (4.16) and (4.17). From the above analysis, it is clear that all simulation results are well consistent with the theoretical analysis.

4.4 Three-Phase Nine-Arm Dual-Input Single-Output AC–DC Converter

4.4.1 Basic Structure and Operating Principle

If the input voltage source is in three-phase form, then there should be three phase units (u, v, and w) for the dual-input single-output AC–DC converter. The corresponding converter is shown in Figure 4.7, named the three-phase nine-arm dual-input single-output AC–DC converter [3].

Similar to Section 4.3.1, the voltages between different phase units and the switching-arm voltages have the following relationship:

$$\begin{cases} u_{\text{u1}} - u_{\text{v1}} = \dfrac{1}{2}[(u_{\text{Lu}} + u_{\text{Mu}} - u_{\text{Uu}}) - (u_{\text{Lv}} + u_{\text{Mv}} - u_{\text{Uv}})] \\[2mm] u_{\text{u2}} - u_{\text{v2}} = \dfrac{1}{2}[(u_{\text{Lu}} - u_{\text{Mu}} - u_{\text{Uu}}) - (u_{\text{Lv}} - u_{\text{Mv}} - u_{\text{Uv}})] \end{cases} \tag{4.22}$$

$$\begin{cases} u_{\text{v1}} - u_{\text{w1}} = \dfrac{1}{2}[(u_{\text{Lv}} + u_{\text{Mv}} - u_{\text{Uv}}) - (u_{\text{Lw}} + u_{\text{Mw}} - u_{\text{Uw}})] \\[2mm] u_{\text{v2}} - u_{\text{w2}} = \dfrac{1}{2}[(u_{\text{Lv}} - u_{\text{Mv}} - u_{\text{Uv}}) - (u_{\text{Lw}} - u_{\text{Mw}} - u_{\text{Uw}})] \end{cases} \tag{4.23}$$

$$\begin{cases} u_{\text{w1}} - u_{\text{u1}} = \dfrac{1}{2}[(u_{\text{Lw}} + u_{\text{Mw}} - u_{\text{Uw}}) - (u_{\text{Lu}} + u_{\text{Mu}} - u_{\text{Uu}})] \\[2mm] u_{\text{w2}} - u_{\text{u2}} = \dfrac{1}{2}[(u_{\text{Lw}} - u_{\text{Mw}} - u_{\text{Uw}}) - (u_{\text{Lu}} - u_{\text{Mu}} - u_{\text{Uu}})] \end{cases} \tag{4.24}$$

4.4.2 Control Scheme

As the fundamental component of the voltage between different phase units is equal to the line voltage of the three-phase input, similar to Eq. (4.14), the reference signals for controlling the switching arm voltages can be defined by

$$\begin{cases} \dfrac{(u_{\text{Lu}} + u_{\text{Mu}} - u_{\text{Uu}}) - (u_{\text{Lv}} + u_{\text{Mv}} - u_{\text{Uv}})}{U_\text{O}} = \dfrac{2u_{\text{1u}} - 2u_{\text{1v}}}{U_\text{O}} = u_{\text{Ref1u}} - u_{\text{Ref1v}} \\[4mm] \dfrac{(u_{\text{Lu}} - u_{\text{Mu}} - u_{\text{Uu}}) - (u_{\text{Lv}} - u_{\text{Mv}} - u_{\text{Uv}})}{U_\text{O}} = \dfrac{2u_{\text{2u}} - 2u_{\text{2v}}}{U_\text{O}} = u_{\text{Ref2u}} - u_{\text{Ref2v}} \end{cases}$$

$$\tag{4.25}$$

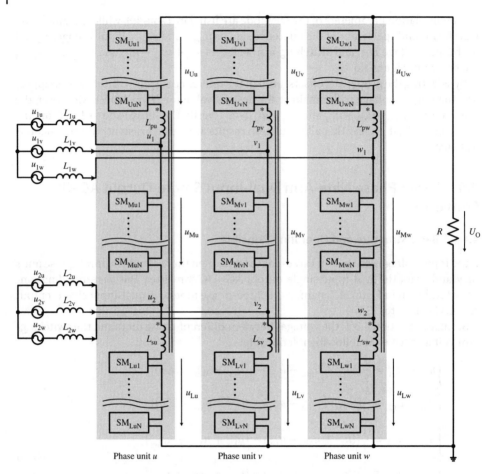

Figure 4.7 Three-phase nine-arm dual-input single-output AC–DC converter.

$$\begin{cases} \dfrac{(u_{\mathrm{Lv}} + u_{\mathrm{Mv}} - u_{\mathrm{Uv}}) - (u_{\mathrm{Lw}} + u_{\mathrm{Mw}} - u_{\mathrm{Uw}})}{U_{\mathrm{O}}} = \dfrac{2u_{1\mathrm{v}} - 2u_{1\mathrm{w}}}{U_{\mathrm{O}}} = u_{\mathrm{Ref1v}} - u_{\mathrm{Ref1w}} \\[4mm] \dfrac{(u_{\mathrm{Lv}} - u_{\mathrm{Mv}} - u_{\mathrm{Uv}}) - (u_{\mathrm{Lw}} - u_{\mathrm{Mw}} - u_{\mathrm{Uw}})}{U_{\mathrm{O}}} = \dfrac{2u_{2\mathrm{v}} - 2u_{2\mathrm{w}}}{U_{\mathrm{O}}} = u_{\mathrm{Ref2v}} - u_{\mathrm{Ref2w}} \end{cases}$$

$$(4.26)$$

$$\begin{cases} \dfrac{(u_{\mathrm{Lw}} + u_{\mathrm{Mw}} - u_{\mathrm{Uw}}) - (u_{\mathrm{Lu}} + u_{\mathrm{Mu}} - u_{\mathrm{Uu}})}{U_{\mathrm{O}}} = \dfrac{2u_{1\mathrm{w}} - 2u_{1\mathrm{u}}}{U_{\mathrm{O}}} = u_{\mathrm{Ref1w}} - u_{\mathrm{Ref1u}} \\[4mm] \dfrac{(u_{\mathrm{Lw}} - u_{\mathrm{Mw}} - u_{\mathrm{Uw}}) - (u_{\mathrm{Lu}} - u_{\mathrm{Mu}} - u_{\mathrm{Uu}})}{U_{\mathrm{O}}} = \dfrac{2u_{2\mathrm{w}} - 2u_{2\mathrm{u}}}{U_{\mathrm{O}}} = u_{\mathrm{Ref2w}} - u_{\mathrm{Ref2u}} \end{cases}$$

$$(4.27)$$

Therefore, using the CPS-SPWM scheme presented before to control the switching-arm voltage, the reference signals u_{Ref1j} and u_{Ref2j}, which correspond to the dual three-phase input voltages, are determined by

$$u_{\text{Ref1j}}(t) = M_1 \sin(2\pi f_1 t + \varphi_1 + \phi_j) + U_{\text{os1}} \tag{4.28}$$

$$u_{\text{Ref2j}}(t) = M_2 \sin(2\pi f_2 t + \varphi_2 + \phi_j) + U_{\text{os2}} \tag{4.29}$$

where $j = $ u, v, w and $\phi_u = 0$, $\phi_v = -\frac{2\pi}{3}$, $\phi_w = \frac{2\pi}{3}$. $M_1 = \frac{2U_1}{U_O}$ and $M_2 = \frac{2U_2}{U_O}$ are modulated ratios, while U_1 and U_2 are phase voltage amplitudes, f_1 and f_2 are frequencies, φ_1 and φ_2 are phase shifts of the dual input AC voltage sources, respectively. U_{os1} and U_{os2} are DC offsets.

The relationship between modulating ratios and DC offset is consistent with that in Section 4.3.2, that is, Eq. (4.18) or Eq. (4.20). However, the output voltage value will be in the range of

$$U_O \geq \begin{cases} 2(U_1 + U_2) & \text{except for EF}' \text{ mode} \\ 2 \times \max(U_1, U_2) & \text{only for EF}' \text{ mode} \end{cases} \tag{4.30}$$

4.4.3 Performance Verification

Based on Figure 4.7, a converter with $N = 4$ was built to verify the performance of the proposed three-phase nine-arm dual-input single-output AC–DC converter. The simulation parameters are listed in Table 4.4, while the converter will operate in the DF mode. The simulation waveforms and spectrums of output voltage U_O, dual input voltages (u_{1u}, u_{2u}), reference signals (u_{Ref1u}, u_{Ref2u}) for phase unit u, and the switching arm voltage combinations ($u_{\text{Lu}} + u_{\text{Mu}} - u_{\text{Uu}}$, $u_{\text{Lu}} - u_{\text{Mu}} - u_{\text{Uu}}$) of phase unit u, are shown in Figure 4.8. It is found that the output voltage remains at the desired value, and the spectrums of $u_{\text{Lu}} + u_{\text{Mu}} - u_{\text{Uu}}$ and $u_{\text{Lu}} - u_{\text{Mu}} - u_{\text{Uu}}$ agree with those of the reference signals u_{reflu} and u_{ref2u}, which proves that the proposed topology and the corresponding control scheme are feasible.

4.5 Single-Phase M-Arm Multiple-Input Single-Output AC–DC Converter

4.5.1 Basic Structure and Operating Principle

If more than two AC inputs need to be converted to one DC output at the same time, similar to Section 3.5, then the single-phase multiple-input single-output AC–DC converter can be constructed by replacing the middle switching arm in Figure 4.1 by

Table 4.4 Parameters of the three-phase nine-arm dual-input single-output AC–DC converter.

U_O (V)	U_1 (V)	M_1	f_1 (Hz)	φ_1 (rad)	U_{os1} (V)	U_2 (V)	M_2	f_2 (Hz)	φ_2 (rad)	U_{os2} (V)
200	60	0.6	50	0	0.4	40	0.4	100	0	−0.6

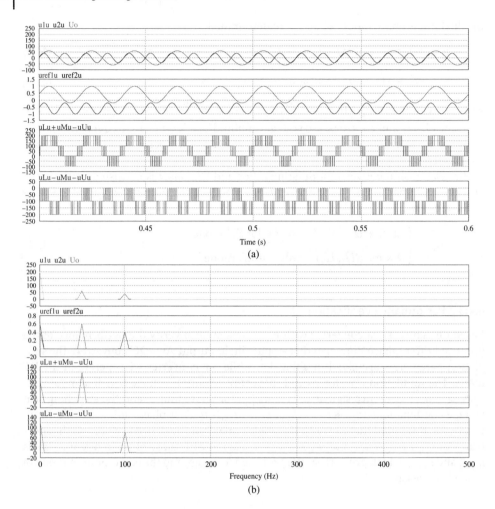

Figure 4.8 Simulation waveforms of the proposed three-phase nine-arm dual-input single-output AC–DC converter in the DF mode. (a) Waveforms. (b) Spectrums.

several switching arms connected in series [4]. As shown in Figure 4.9, there are M switching arms and M capacitors in total, M − 1 single-phase AC voltage sources are connected between the switching arms and the capacitors.

From Figure 4.9, the voltages of u_{a1} and $u_{a(M-1)}$ can be defined by

$$\begin{cases} u_{a1} = U_O - u_{A1} - u_{Lp} \\ u_{a(M-1)} = u_{AM} + u_{Ls} \end{cases} \tag{4.31}$$

Assume that the primary inductance L_p, the secondary inductance L_s, and the mutual inductance M_L of the coupled inductor are identical, then $u_{Lp} = u_{Ls}$. We have

$$u_{a1} + u_{a(M-1)} = U_O - u_{A1} + u_{AM} \tag{4.32}$$

The difference between u_{a1} and $u_{a(M-1)}$ is equal to

$$u_{a1} - u_{a(M-1)} = u_{A2} + u_{A3} + \cdots + u_{A(M-1)} \tag{4.33}$$

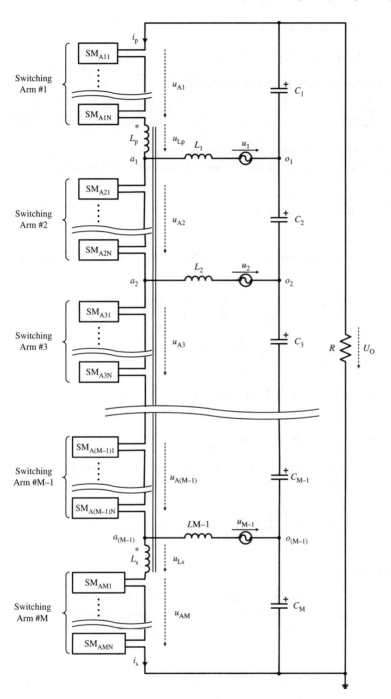

Figure 4.9 Single-phase M-arm multiple-input single-output AC–DC converter.

Thus, the voltage at the joint point a_1 is

$$u_{a1} = \frac{1}{2}(U_O - u_{A1} + u_{A2} + u_{A3} + \cdots + u_{AM}) \tag{4.34}$$

and the other joint-point voltage can be determined by

$$u_{ax} = u_{a1} - \sum_{i=2}^{x} u_{Ai} \tag{4.35}$$

where $x = 2, 3,\ldots,M-1$, u_{Ai} is the voltage of the ith switching arm.

Substituting Eq. (4.34) into Eq. (4.35), the kth joint-point voltage in Figure 4.9 is

$$u_{ak} = \frac{1}{2}\left(U_O - \sum_{i=1}^{k} u_{Ai} + \sum_{j=k+1}^{M} u_{Aj}\right) \tag{4.36}$$

where $k = 1, 2,\ldots, M-1$.

According to Kirchhoff's voltage law, the voltage of the kth input branch at steady state is expressed by

$$u_{akok} = u_{ak} - u_{ok} = \frac{1}{2}\left(U_O - \sum_{i=1}^{k} u_{Ai} + \sum_{j=k+1}^{M} u_{Aj}\right) - \frac{M-k}{M}U_O$$

$$= \frac{1}{2}\left(\frac{2k-M}{M}U_O - \sum_{i=1}^{k} u_{Ai} + \sum_{j=k+1}^{M} u_{Aj}\right) \tag{4.37}$$

It is found that the output voltage is determined by

$$U_O = \frac{M}{2k-M}\left(2u_{akok} + \sum_{i=1}^{k} u_{Ai} - \sum_{j=k+1}^{M} u_{Aj}\right) \tag{4.38}$$

Therefore, the desired output voltage can be obtained by regulating the switching-arm voltages, which should satisfy

$$\sum_{j=k+1}^{M} u_{Aj} - \sum_{i=1}^{k} u_{Ai} = 2u_{akok} + \frac{M-2k}{M}U_O \tag{4.39}$$

4.5.2 Control Scheme

If the CPS-SPWM scheme is applied to the proposed converter in Figure 4.9, then the control schematic in Figure 3.12 is also available. As the fundamental voltage component u_{akok} is equal to the input voltage u_k, the reference signals for Eq. (4.39) can be determined by

$$u_{Refk} = \frac{1}{U_O}\left(\sum_{j=k+1}^{M} u_{Aj} - \sum_{i=1}^{k} u_{Ai}\right) = M_k \sin(2\pi f_k t + \varphi_k) + U_{osk} \tag{4.40}$$

where $M_k = \frac{2U_k}{U_O}$ is the modulated ratio, while U_k, f_k, and φ_k are the magnitude, frequency, and phase angle shift of the kth AC input voltage source, respectively. The DC offset is equal to $U_{osk} = \frac{M-2k}{M}$, based on Eq. (4.39).

In order to avoid overlap of the reference signals, $u_{\text{Ref}(k-1)} \geq u_{\text{Refk}}$ should be satisfied, then we have

$$
\begin{cases}
-M_1 + \dfrac{M-2}{M} \geq M_2 + \dfrac{M-2\times 2}{M} \\[2mm]
-M_2 + \dfrac{M-2\times 2}{M} \geq M_3 + \dfrac{M-2\times 3}{M} \\[2mm]
\cdots \\[1mm]
-M_{k-1} + \dfrac{M-2(k-1)}{M} \geq M_k + \dfrac{M-2k}{M} \qquad\qquad \text{except for the EF}' \text{ mode}\\[2mm]
\cdots \\[1mm]
-M_{M-2} + \dfrac{M-2(M-2)}{M} \geq M_{M-1} + \dfrac{M-2(M-1)}{M}
\end{cases}
\tag{4.41}
$$

By combining all the sub-equations in Eq. (4.41) together, the following equation is obtained:

$$
-\sum_{i=1}^{M-2} M_i - \frac{2}{M} \geq \sum_{j=2}^{M-1} M_j - \frac{2(M-1)}{M}
\tag{4.42}
$$

Then the range of the output voltage will be

$$
U_O \geq \frac{M}{M-2}\left(U_1 + 2\sum_{x=2}^{M-2} U_x + U_{M-1} \right)
\tag{4.43}
$$

If the proposed converter operates in the EF' mode, which means that all input AC voltages have the same frequency and phase shift, then

$$
1 \geq \left(M_1 + \frac{M-2}{M} \right) \geq \left(M_2 + \frac{M-2\times 2}{M} \right)
$$
$$
\geq \cdots \geq \left(M_{k-1} + \frac{M-2(k-1)}{M} \right) \geq \left(M_k + \frac{M-2k}{M} \right)
$$
$$
\geq \cdots \geq \left(M_{M-2} + \frac{M-2(M-2)}{M} \right) \geq \left(M_{M-1} + \frac{M-2(M-1)}{M} \right)
\tag{4.44a}
$$

and

$$
\left(-M_1 + \frac{M-2}{M} \right) \geq \left(-M_2 + \frac{M-2\times 2}{M} \right)
$$
$$
\geq \cdots \geq \left(-M_{k-1} + \frac{M-2(k-1)}{M} \right) \geq \left(-M_k + \frac{M-2k}{M} \right)
$$
$$
\geq \cdots \geq \left(-M_{M-2} + \frac{M-2(M-2)}{M} \right) \geq \left(-M_{M-1} + \frac{M-2(M-1)}{M} \right) \geq -1
\tag{4.44b}
$$

Then we have

$$
|M_k - M_{k-1}| \leq \frac{2}{M}
\tag{4.45}
$$

Table 4.5 Parameters of the single-phase four-arm multiple-input single-output AC–DC converter.

Mode	M	N	Input voltages (V)			Output voltage U_0 (V)
			$k = 1$	$k = 2$	$k = 3$	
DF	4	4	$50 \sin 100\pi t$	$50 \sin \left(100\pi t - \dfrac{2\pi}{3}\right)$	$50 \sin \left(100\pi t + \dfrac{2\pi}{3}\right)$	800
EF′	4	4	$100 \sin 100\pi t$	$300 \sin 100\pi t$	$200 \sin 100\pi t$	800

which means that the output voltage will be in the range of

$$U_O \geq \max(M \cdot |U_k - U_{k-1}|) \tag{4.46}$$

When $M = 3$, there are two input AC sources, the same as in Figure 4.1. It is found that Eqs. (4.43) and (4.46) are consistent with Eq. (4.10) in Section 4.2.

4.5.3 Performance Verification

In this section, a four-arm three-input simulation prototype is used as an example to verify the performance of the proposed converter in Figure 4.9; the simulation parameters are listed in Table 4.5.

When the simulation prototype is operating in the DF mode, the simulation waveforms of three input voltages, the output voltage, and the reference signals are shown in Figure 4.10a, where the output voltage remains at the desired value. The voltages of $\sum_{j=k+1}^{M} u_{Aj} - \sum_{i=1}^{k} u_{Ai}$ in which $k = 1$, 2, 3 and their spectrums are illustrated in Figure 4.10b,c, respectively. It is found that the average values and spectrums of $\sum_{j=k+1}^{M} u_{Aj} - \sum_{i=1}^{k} u_{Ai}$ agree with Eq. (4.39).

When the simulation prototype is operating in the EF′ mode, the corresponding simulation waveforms are shown in Figure 4.11. It is clear that the definitions of the reference signals and the output voltage are feasible.

4.6 Single-Phase 2M-Arm Multiple-Input Single-Output AC–DC Converter

4.6.1 Basic Structure and Operating Principle

In order to avoid using high-voltage capacitors in the topology, the single-phase 2M-arm multiple-input single-output AC–DC converter can be constructed by replacing the series capacitors in Figure 4.9 with switching arms [5]. As shown in Figure 4.12, there are two general phase units in the proposed 2M-arm AC–DC converter, which consists of M switching arms and one coupled inductor.

Similar to Section 4.5.1, the kth joint-point voltage of the general phase unit is determined by

$$u_{xk} = \frac{1}{2}\left(U_O - \sum_{i=1}^{k} u_{Xi} + \sum_{j=k+1}^{M} u_{Xj}\right) \tag{4.47}$$

where $k = 1, 2,..., M - 1$. If $x = a$ then $X = A$, if $x = b$ then $X = B$.

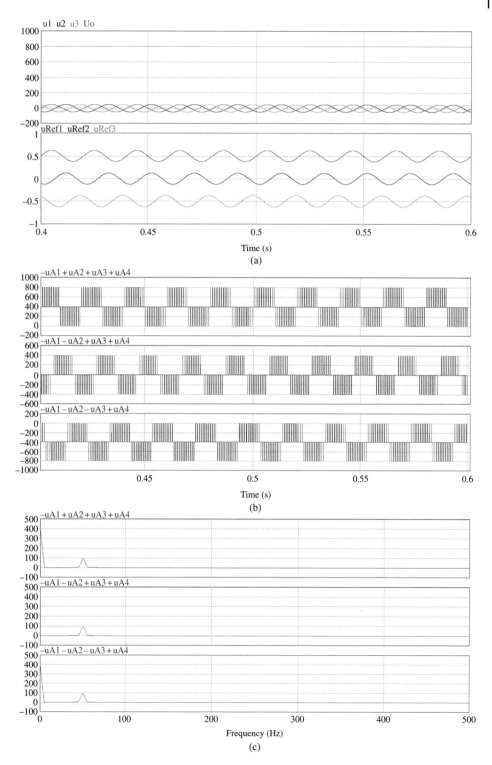

Figure 4.10 Simulation waveforms of the proposed single-phase four-arm three-input single-output AC–DC converter in the DF mode. (a) Input voltages, output voltage, and reference voltages. (b) Switching-arm voltage combinations. (c) Spectrums of switching-arm voltage combinations.

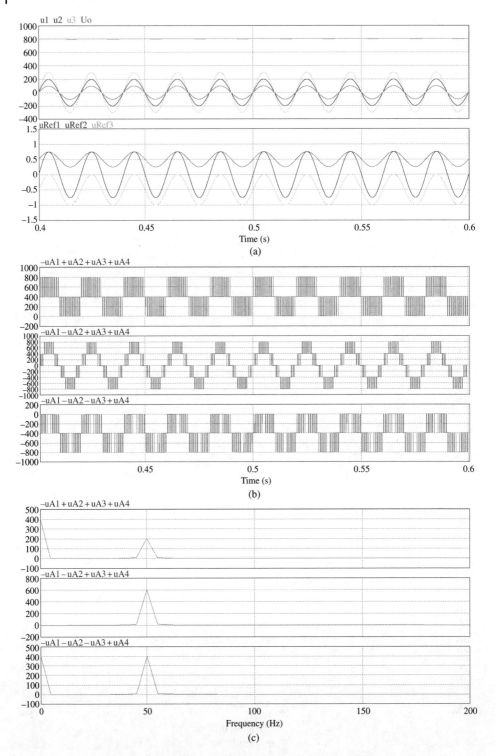

Figure 4.11 Simulation waveforms of the proposed single-phase four-arm three-input single-output AC–DC converter in the EF′ mode. (a) Input voltages, output voltage, and reference voltages. (b) Switching-arm voltage combinations. (c) Spectrums of switching-arm voltage combinations.

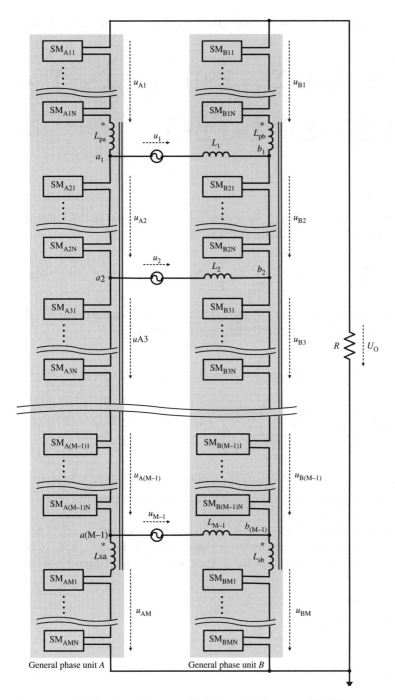

Figure 4.12 Single-phase 2M-arm multiple-input single-output AC–DC converter.

According to Kirchhoff's voltage law, the voltage of the kth input branch at steady state is expressed by

$$u_{akbk} = u_{ak} - u_{bk} = \frac{1}{2}\left(U_O - \sum_{i=1}^{k} u_{Ai} + \sum_{j=k+1}^{M} u_{Aj}\right) - \frac{1}{2}\left(U_O - \sum_{i=1}^{k} u_{Bi} + \sum_{j=k+1}^{M} u_{Bj}\right)$$

$$= \frac{1}{2}\left(-\sum_{i=1}^{k} u_{Ai} + \sum_{j=k+1}^{M} u_{Aj} + \sum_{i=1}^{k} u_{Bi} - \sum_{j=k+1}^{M} u_{Bj}\right) \tag{4.48}$$

4.6.2 Control Scheme

As the fundamental component of u_{akbk} or $u_{ak} - u_{bk}$ should be equal to the input AC voltage u_k in the steady state, Eq. (4.48) turns out to be

$$\left(-\sum_{i=1}^{k} u_{Ai} + \sum_{j=k+1}^{M} u_{Aj} + \sum_{i=1}^{k} u_{Bi} - \sum_{j=k+1}^{M} u_{Bj}\right) = 2u_k \tag{4.49}$$

In order to obtain a constant output DC voltage, the reference signals to control the switching arm voltages of different general phase units are defined by

$$\begin{cases} u_{Re\,fak} = \dfrac{1}{U_O}\left(-\sum_{i=1}^{k} u_{Ai} + \sum_{j=k+1}^{M} u_{Aj}\right) = \dfrac{u_k}{U_O} + U_{osk} = M_k \sin(2\pi f_k t + \varphi_k) + U_{osk} \\[4mm] u_{Refbk} = \dfrac{1}{U_O}\left(-\sum_{i=1}^{k} u_{Bi} + \sum_{j=k+1}^{M} u_{Bj}\right) = -\dfrac{u_k}{U_O} + U_{osk} = -M_k \sin(2\pi f_k t + \varphi_k) + U_{osk} \end{cases}$$

$$\tag{4.50}$$

where $M_k = \frac{U_k}{U_O}$ and U_{osk} are the modulated ratio and the DC offset of the kth reference signals, respectively. U_k, f_k, and φ_k are amplitude, frequency, and phase shift of the kth input voltage, respectively.

If the CPS-SPWM scheme is selected to control the proposed AC–DC converter, then the reference signals should meet the requirements of $u_{Refa(k-1)} \geq u_{Refak}$ and $u_{Refb(k-1)} \geq u_{Refbk}$ in order to avoid their overlap. In the case of DF and EF modes, we have

$$\begin{cases} 1 \geq (M_1 + U_{os1}) \\ (-M_1 + U_{os1}) \geq (M_2 + U_{os2}) \\ \dots \\ (-M_{k-1} + U_{os(k-1)}) \geq (M_k + U_{osk}) \\ \dots \\ (-M_{M-2} + U_{os(M-2)}) \geq (M_{M-1} + U_{os(M-1)}) \\ (-M_{M-1} + U_{os(M-1)}) \geq -1 \end{cases} \tag{4.51}$$

Thus, the DC offsets should satisfy

$$
\begin{cases}
U_{os1} \leq 1 - M_1 \\
U_{os2} \leq U_{os1} - M_1 - M_2 \\
\quad \cdots \\
U_{osk} \leq U_{os(k-1)} - M_{k-1} - M_k \\
\quad \cdots \\
U_{os(M-1)} \leq U_{os(M-2)} - M_{M-2} - M_{M-1} \leq 1 - M_{M-1}
\end{cases}
\tag{4.52}
$$

By combining all the sub-equations in Eq. (4.51) together, the following equation is obtained:

$$
\sum_{k=1}^{M-1} M_k \leq 1
\tag{4.53}
$$

and the range of the output voltage for the DF and EF modes will be

$$
U_O \geq \sum_{k=1}^{M-1} U_k
\tag{4.54}
$$

Therefore, M_k can be obtained and U_{osk} can be selected based on Eq. (4.52) when the DC output voltage U_O is determined.

For the case of the EF′ mode, the reference signals have the following relationship:

$$
\begin{cases}
1 \geq (M_1 + U_{os1}) \quad \geq (M_2 + U_{os2}) \geq \cdots \geq (M_{k-1} + U_{os(k-1)}) \geq (M_k + U_{osk}) \\
\qquad\qquad\qquad \geq \cdots \geq (M_{M-2} + U_{os(M-2)}) \geq (M_{M-1} + U_{os(M-1)}) \\
(-M_1 + U_{os1}) \quad \geq (-M_2 + U_{os2}) \geq \cdots \geq (-M_{k-1} + U_{os(k-1)}) \geq (-M_k + U_{osk}) \\
\qquad\qquad\qquad \geq \cdots \geq (-M_{M-2} + U_{os(M-2)}) \geq (-M_{M-1} + U_{os(M-1)}) \geq -1
\end{cases}
\tag{4.55}
$$

Then the DC offsets of the nearby references should satisfy

$$
U_{os(k-1)} - U_{osk} \geq |M_k - M_{k-1}|
\tag{4.56}
$$

Because the modulated ratio is normally determined by $M_k < 1$, the output voltage will be larger than any of the input voltage amplitudes, that is

$$
U_O > \max(U_1, U_2, \ldots, U_{M-1})
\tag{4.57}
$$

When $M = 3$, there are two input AC sources, the same as in Figure 4.4. It is found that Eqs. (4.54) and (4.57) are consistent with the analysis in Section 4.3.2.

4.6.3 Performance Verification

The performance of the proposed 2M-arm AC–DC converter in Figure 4.12 will be verified by an eight-arm three-input simulation prototype with $M = N = 4$. The simulation parameters are listed in Table 4.6, in which the desired output voltage is $U_O = 800\,\text{V}$, and the DC offsets are selected based on Eq. (4.52).

Table 4.6 Parameters of the single-phase eight-arm three-input single-output AC–DC converter.

Mode	k	1	2	3
DF	Input voltages u_k	$100\sin 100\pi t$	$100\sin\left(100\pi t - \frac{2\pi}{3}\right)$	$100\sin\left(100\pi t + \frac{2\pi}{3}\right)$
	Reference signals u_{Refak}	$\frac{1}{8}\sin 100\pi t + \frac{1}{2}$	$\frac{1}{8}\sin\left(100\pi t - \frac{2\pi}{3}\right)$	$\frac{1}{8}\sin\left(100\pi t + \frac{2\pi}{3}\right) - \frac{1}{2}$
EF'	Input voltages u_k	$200\sin 100\pi t$	$400\sin 100\pi t$	$600\sin 100\pi t$
	Reference signals u_{Refak}	$\frac{1}{4}\sin 100\pi t + \frac{3}{4}$	$\frac{1}{2}\sin 100\pi t$	$\frac{3}{4}\sin 100\pi t - \frac{1}{4}$

When the simulation prototype is operating in the DF mode, the simulation waveforms of three input voltages, the output voltage, and two groups of reference signals are shown in Figure 4.13a, where the output voltage remains at the desired value. The voltages of three input branches and their spectrums are illustrated in Figure 4.13b,c, respectively, which proves the correctness of Eq. (4.48).

When the simulation prototype is operating in the EF' mode, the corresponding simulation waveforms are shown in Figure 4.14. It is clear that the definitions of the reference signals and the output voltage are feasible.

4.7 Three-Phase 3M-Arm Multiple-Input Single-Output AC–DC Converter

4.7.1 Basic Structure and Operating Principle

If the input voltage sources are in three-phase form, then the three-phase multiple-input single-output AC–DC converter can be constructed by adding one general phase unit to the single-phase 2M-arm AC–DC converter. Thus, there are 3M arms in the proposed three-phase multiple-input single-output AC–DC converter, which is shown in Figure 4.15 [6].

Similar to Section 4.6.1, the voltage between the switching arms at steady state can be expressed by

$$u_{kx} - u_{ky} = \frac{1}{2}\left(-\sum_{i=1}^{k} u_{Xi} + \sum_{j=k+1}^{M} u_{Xj} + \sum_{i=1}^{k} u_{Yi} - \sum_{j=k+1}^{M} u_{Yj} \right) \tag{4.58}$$

where $k = 1, 2, \ldots, M - 1$, $x, y = \{u, v, w\}$, and X, Y $= \{U, V, W\}$.

4.7.2 Control Scheme

Normally, the three-phase input voltage is in the form

$$u_{kx}(t) = U_k \sin(2\pi f_k t + \varphi_k + \phi_x) \tag{4.59}$$

where $\phi_u = 0$, $\phi_v = -\frac{2\pi}{3}$, $\phi_w = \frac{2\pi}{3}$, U_k, f_k, and φ_k are the amplitude, frequency, and phase shift, respectively.

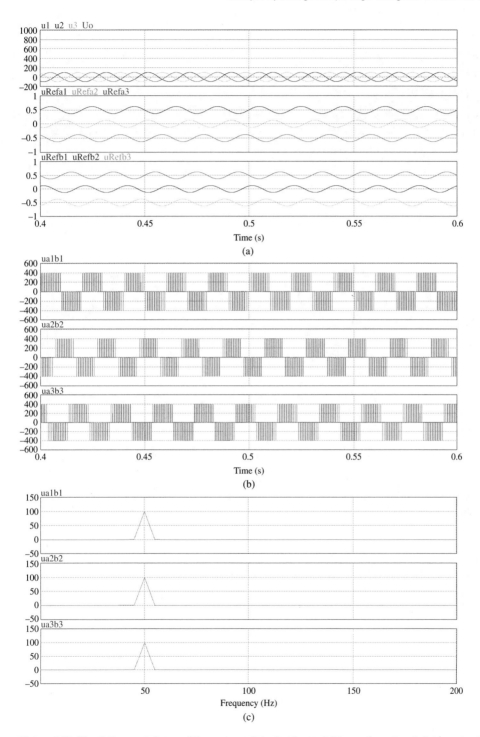

Figure 4.13 Simulation waveforms of the proposed single-phase eight-arm three-input single-output AC–DC converter in the DF mode. (a) Input voltages, output voltage, and reference voltages. (b) Voltages of three input branches. (c) Spectrums of (b).

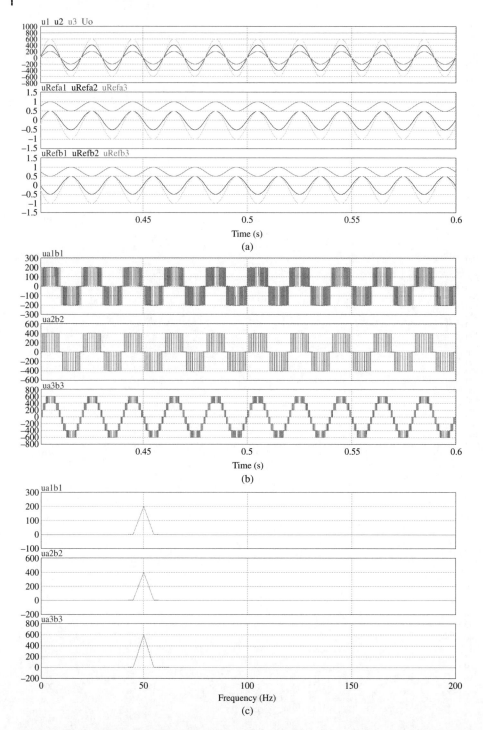

Figure 4.14 Simulation waveforms of the proposed single-phase eight-arm three-input single-output AC–DC converter in the EF′ mode. (a) Input voltages, output voltage, and reference voltages. (b) Switching arm voltage combinations. (c) Spectrums of (b).

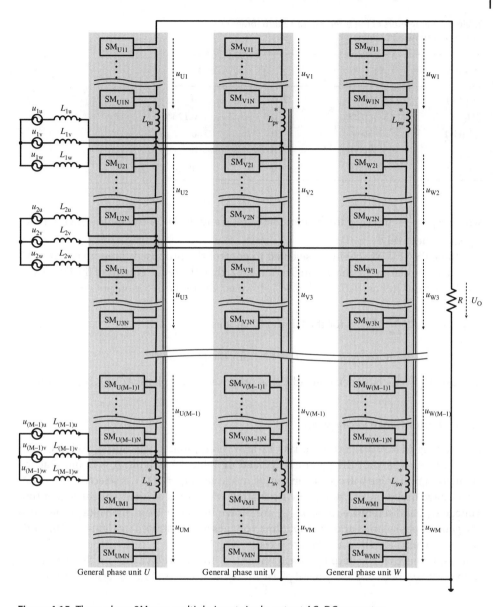

Figure 4.15 Three-phase 3M-arm multiple-input single-output AC–DC converter.

If a constant DC output voltage U_O is wanted, the reference signals to regulate the switching-arm voltages should be determined based on Eq. (4.58), that is

$$
\begin{cases}
u_{\mathrm{Re\,} fkx} = \dfrac{1}{U_O}\left(-\sum_{i=1}^{k} u_{Xi} + \sum_{j=k+1}^{M} u_{Xj}\right) = \dfrac{2u_{kx}}{U_O} + U_{osk} \\[2mm]
\qquad = M_k \sin(2\pi f_k t + \varphi_k + \phi_x) + U_{osk} \\[3mm]
u_{\mathrm{Re\,} fky} = \dfrac{1}{U_O}\left(-\sum_{i=1}^{k} u_{Yi} + \sum_{j=k+1}^{M} u_{Yj}\right) = \dfrac{2u_{ky}}{U_O} + U_{osk} \\[2mm]
\qquad = M_k \sin(2\pi f_k t + \varphi_k + \phi_y) + U_{osk}
\end{cases}
\tag{4.60}
$$

Table 4.7 Parameters of the three-phase 12-arm three-input single-output AC–DC converter.

k	1	2	3
U_k (V)	100	200	50
M_k	0.25	0.5	0.125
f_k (Hz)	50	50	60
$\varphi_{k\,(rad)}$	0	0	0
U_{osk} (V)	0.75	0	−0.75

where $M_k = \frac{2U_k}{U_O}$ and U_{osk} are the modulated ratio and the DC offset for the kth reference signal, respectively.

If the CPS-SPWM scheme is selected to control the proposed three-phase AC–DC converter, similar to the analysis in Section 4.6.2, then the requirements of the DC offsets are the same as in Eqs. (4.52) and (4.56). However, the range of the output voltage will be

$$U_O \geq 2 \sum_{k=1}^{M-1} U_k \qquad \text{for the DF and EF modes} \tag{4.61}$$

$$U_O > 2 \times \max(U_1, U_2, \ldots, U_{M-1}) \qquad \text{for the EF}' \text{ mode} \tag{4.62}$$

because $M_k = \frac{2U_k}{U_O}$ has been defined in the proposed three-phase AC–DC converter.

4.7.3 Performance Verification

By selecting $M = N = 4$, a three-phase 12-arm three-input single-output AC–DC simulation prototype with the parameters listed in Table 4.7 is built. It is found that the simulation prototype will operate in the DF mode and $U_O = 800$ V is selected as the output voltage based on Eq. (4.61). The simulation waveforms of input voltage u_{ku}, output voltage U_O, and reference signal u_{Refuk} are shown in Figure 4.16a, the voltages between the general phase units and the input line voltages are illustrated in Figure 4.16b, while their spectrums are shown in Figure 4.16c.

It is obvious that the output voltage remains at the desired value, and the spectrums of the voltages between the general phase units and the input line voltages are the same.

4.8 Summary

In this chapter, a group of multiple-input single-output high-voltage AC–DC converters have been proposed, in which two or more single-phase/three-phase AC input sources with identical or different frequencies or phase shifts can be converted to one DC output. The proposed converter has the advantage of a simple structure and fewer components, which can be applied to connect different AC voltages and is suitable for multi-terminal HVDC systems (e.g. wind turbine generator with open winding structure).

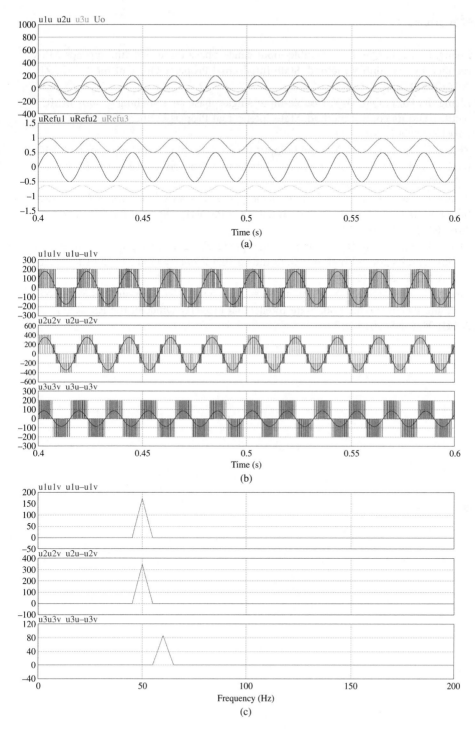

Figure 4.16 Simulation waveforms of the proposed three-phase 12-arm three-input single-output AC–DC converter in the DF mode. (a) Input voltages, output voltage, and reference voltages. (b) Voltages between general phase units and input line voltages. (c) Spectrums of (b).

References

1 Zhang, B., Fu, J., Qiu, D. Y. Double-input single-phase three switching-groups MMC rectifier and its control method. State Intellectual Property Office of the P.R.C., ZL 201410043022.6, 2016.10.5.

2 Zhang, B., Fu, J., Qiu, D. Y. Double-input single-phase six switching-groups MMC rectifier and its control method. State Intellectual Property Office of the P.R.C., ZL 201410042758.1, 2017.1.11.

3 Zhang, B., Fu, J., Qiu, D. Y. Double-input three-phase nine switching-groups MMC rectifier and its control method. State Intellectual Property Office of the P.R.C., ZL 201410042977.x, 2016.7.6.

4 Zhang, B., Fu, J., Qiu, D. Y. N-input single-phase N+1 switching-groups MMC rectifier and its control method. State Intellectual Property Office of the P.R.C., ZL 201410042775.5, 2016.7.6.

5 Zhang, B., Fu, J., Qiu, D. Y. N-input single-phase 2N+2 switching-groups MMC rectifier and its control method. State Intellectual Property Office of the P.R.C., ZL 201410042800.X, 2017.1.11.

6 Zhang, B., Fu, J., Qiu, D. Y. N-input three-phase 3N+3 switching-groups MMC rectifier and its control method. State Intellectual Property Office of the P.R.C., ZL 201410042836.8, 2016.4.13.

5

Multiple-Input Multiple-Output High-voltage AC–AC Converters

5.1 Introduction

A multiple-input multiple-output high-voltage AC–AC converter will be useful in situations with several AC inputs or outputs. For example, if there are many wind turbine generators in a large wind farm, whose output power should be connected together and sent to the grid. In the existing schemes, each wind turbine generator needs a converter to regulate its output power. By using the multiple-input multiple-output AC–AC converter, several wind turbine generators can share one converter, which results in reduced switching components and simplified system structure.

Based on the multi-terminal DC–AC and AC–DC converters presented in Chapters 3 and 4, it is found that both AC voltage source and AC load can be connected between the switching arms. Thus, the multiple-input multiple-output high-voltage AC–AC converter can be obtained by using some branches as AC voltage sources and others as AC loads. In this chapter, the single-input single-output AC–AC converter, which is a special case of the multi-terminal AC–AC converter, will be introduced first to describe the operating principle of the proposed structure, then the single-phase and three-phase multiple-input multiple-output converters will be discussed, respectively. The carrier phase-shifted sinusoidal pulse-width modulation (CPS-SPWM) scheme has been applied to the proposed converter, and the simulation waveforms will be provided to verify the feasibility of the proposed high-voltage AC–AC converters.

5.2 Single-Phase Single-Input Single-Output AC–AC Converter

5.2.1 Basic Structure and Operating Principle

In the single-phase six-arm dual-input single-output AC–DC converter shown in Figure 4.4, there are two AC inputs and one DC output. If one of the AC inputs, for example, u_2, is replaced by a load, and the DC output load is removed, then the single-input single-output AC–AC converter can be obtained [1]. As shown in Figure 5.1, the proposed converter is made up of two phase units, each phase unit has three switching arms and one coupled inductor. The AC input u_i and the AC load output u_o are placed in branch a_1b_1 and a_2b_2 between two phase units, respectively.

Multi-terminal High-voltage Converter, First Edition. Bo Zhang and Dongyuan Qiu.
© 2019 John Wiley & Sons Singapore Pte. Ltd. Published 2019 by John Wiley & Sons Singapore Pte. Ltd.

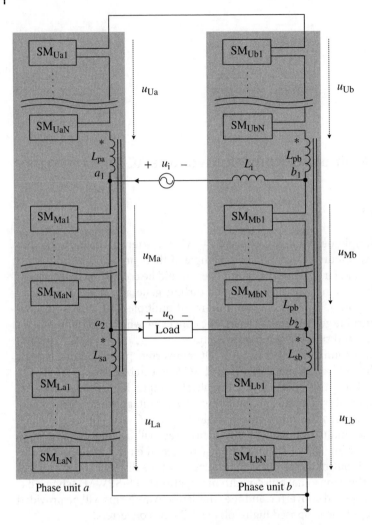

Figure 5.1 Single-phase single-input single-output AC–AC converter.

Similar to the analysis in Section 4.3, the relationships between the input and output voltages and the switching arm voltages are described as follows:

$$u_i = u_{a1} - u_{b1} = \frac{1}{2}[(u_{La} + u_{Ma} - u_{Ua}) - (u_{Lb} + u_{Mb} - u_{Ub})] \tag{5.1}$$

$$u_o = u_{a2} - u_{b2} = \frac{1}{2}[(u_{La} - u_{Ma} - u_{Ua}) - (u_{Lb} - u_{Mb} - u_{Ub})] \tag{5.2}$$

5.2.2 Control Scheme

If the structure of the sub-module (SM) in the switching arm is selected to be the HBSM shown in Figure 3.2, then Figure 5.1 turns into Figure 5.2. When the CPS-SPWM scheme is selected to control the proposed AC–AC converter, as presented in Chapter 3, there

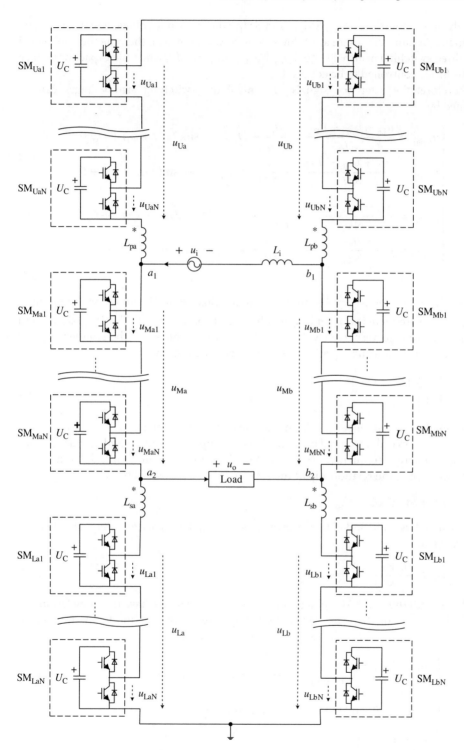

Figure 5.2 Single-phase single-input single-output AC–AC converter with half-bridge sub-module.

is only one SM among SM_{Ui}, SM_{Mi}, and SM_{Li} whose output voltage equals U_C at any moment. Assume that there are N SMs in each switching arm, then the total voltage of the three switching arms will equal $u_{La} + u_{Ma} + u_{Ua} = NU_C$, when the capacitor voltage of the SM is controlled to be constant.

Therefore, the reference signals to control the switching arm voltages can be defined by

$$
\begin{cases}
u_{Refa1} = \dfrac{u_{La} + u_{Ma} - u_{Ua}}{NU_C} = \dfrac{u_i}{NU_C} + O_{si} = M_i \sin(2\pi f_i t + \varphi_i) + O_{si} \\[2mm]
u_{Refb1} = \dfrac{u_{Lb} + u_{Mb} - u_{Ub}}{NU_C} = -\dfrac{u_i}{NU_C} + O_{si} = -M_i \sin(2\pi f_i t + \varphi_i) + O_{si}
\end{cases}
\tag{5.3}
$$

$$
\begin{cases}
u_{Refa2} = \dfrac{u_{La} - u_{Ma} - u_{Ua}}{NU_C} = \dfrac{u_o}{NU_C} + O_{so} = M_o \sin(2\pi f_o t + \varphi_o) + O_{so} \\[2mm]
u_{Refb2} = \dfrac{u_{Lb} - u_{Mb} - u_{Ub}}{NU_C} = -\dfrac{u_o}{NU_C} + O_{so} = -M_o \sin(2\pi f_o t + \varphi_o) + O_{so}
\end{cases}
\tag{5.4}
$$

where $M_i = \dfrac{U_i}{NU_C}$ and $M_o = \dfrac{U_o}{NU_C}$ are modulated ratios, while U_i and U_o are amplitudes, f_i and f_o are frequencies, φ_i and φ_o are phase shifts of the input AC voltage source and output AC voltage, respectively. O_{si} and O_{so} are the DC offsets of the reference signals and satisfy the following equations to avoid over-modulation:

$$
\begin{cases}
0 \le O_{si} \le 1 - M_i \\
0 \ge O_{so} \ge M_o - 1
\end{cases}
\tag{5.5}
$$

Normally, the reference signals should meet the requirements of $u_{Refa1} \ge u_{Refa2}$ and $u_{Refb1} \ge u_{Refb2}$ in order to avoid overlap. When the input AC voltage and output AC voltage have different frequencies (DF mode) or the same frequency with different phase shift (EF mode), the relationship of the two modulated ratios will be

$$
M_i + M_o \le O_{si} - O_{so}
\tag{5.6}
$$

Based on Eqs. (5.5) and (5.6), we have $M_i + M_o \le 1$; the capacitor voltage should be in the range of

$$
U_C \ge \frac{U_i + U_o}{N}
\tag{5.7}
$$

When both the input AC voltage and the output AC voltage have the same frequency and phase shift, or the proposed converter operates in the EF′ mode, the modulated ratio and the DC offset of the reference signals should satisfy

$$
\begin{cases}
M_i + O_{si} \ge M_o + O_{so} \\
-M_i + O_{si} \ge -M_o + O_{so}
\end{cases}
\Rightarrow |M_i - M_o| \le O_{si} - O_{so}
\tag{5.8}
$$

Based on Eqs. (5.5) and (5.8), we have the following equation:

$$
\begin{cases}
M_i + (1 - M_i) \ge M_o + (M_o - 1) \\
-M_i + (1 - M_i) \ge -M_o + (M_o - 1)
\end{cases}
\tag{5.9}
$$

Table 5.1 Parameters of the proposed single-phase single-input single-output AC–AC converter.

Mode	Voltage expression (V)	N	U_C (V)	Reference expression	
DF	$u_i = 100 \sin 100\pi t$	4	50	$u_{Refa1} = 0.5 \sin 100\pi t + 0.5$	$u_{Refb1} = -0.5 \sin 100\pi t + 0.5$
	$u_o = 50 \sin 200\pi t$			$u_{Refa2} = 0.25 \sin 200\pi t - 0.5$	$u_{Refb2} = -0.25 \sin 200\pi t - 0.5$
EF′	$u_i = 160 \sin 100\pi t$			$u_{Refa1} = 0.8 \sin 100\pi t + 0.2$	$u_{Refb1} = -0.8 \sin 100\pi t + 0.2$
	$u_o = 100 \sin 100\pi t$			$u_{Refa2} = 0.4 \sin 100\pi t - 0.5$	$u_{Refb2} = -0.4 \sin 100\pi t - 0.5$

Thus, the capacitor voltage should satisfy

$$U_C \geq \frac{1}{N} \times \max(U_i, U_o) \tag{5.10}$$

5.2.3 Output Voltage Verification

In this section, the proposed converter shown in Figure 5.2 will be verified using the CPS-SPWM scheme in Figure 3.4. In the DF mode, if the input voltage is known as $u_i = 100 \sin 100\pi t$ and the desired output voltage is $u_o = 50 \sin 200\pi t$, then by selecting $N = 4$, the capacitor voltage is controlled to be $U_C = 50$ V based on Eq. (5.7). Because the modulated ratios are $M_i = 0.5$ and $M_o = 0.25$, $O_{si} = 0.5$ and $O_{so} = -0.5$ are chosen based on Eqs. (5.5) and (5.6). The simulation parameters in the EF′ mode can be obtained in a similar way, and all of the simulation parameters are summarized in Table 5.1.

The simulation waveforms in the DF mode are shown in Figure 5.3. It is found that the reference signals defined by Eqs. (5.3) and (5.4) do not overlap, and the total voltage of the switching arms is 200 V, which agrees with $NU_C = 200$ V. The spectrums of the output voltage u_o show that its fundamental component is 50 V at 100 Hz, which is the value we want. The corresponding simulation waveforms in the EF′ mode are shown in Figure 5.4, in which the fundamental component of the output voltage is a scaled-down version of the input voltage. Therefore, the proposed converter can successfully convert one AC input voltage to another AC output voltage.

5.3 Three-Phase Single-Input Single-Output AC–AC Converter

5.3.1 Basic Structure and Operating Principle

If the input and output voltages become three-phase, then one phase unit should be added to Figure 5.1, and the corresponding three-phase AC–AC converter is shown in Figure 5.5 [2].

Similar to Eqs. (5.1) and (5.2), the voltages between different phase units and the switching-arm voltages have the following relationship:

$$\begin{cases} u_{ix} - u_{iy} = u_{1x} - u_{1y} = \frac{1}{2}[(u_{Lx} + u_{Mx} - u_{Ux}) - (u_{Ly} + u_{My} - u_{Uy})] \\ u_{ox} - u_{oy} = u_{2x} - u_{2y} = \frac{1}{2}[(u_{Lx} - u_{Mx} - u_{Ux}) - (u_{Ly} - u_{My} - u_{Uy})] \end{cases} \tag{5.11}$$

where $x, y = $ u, v, w, $x \neq y$.

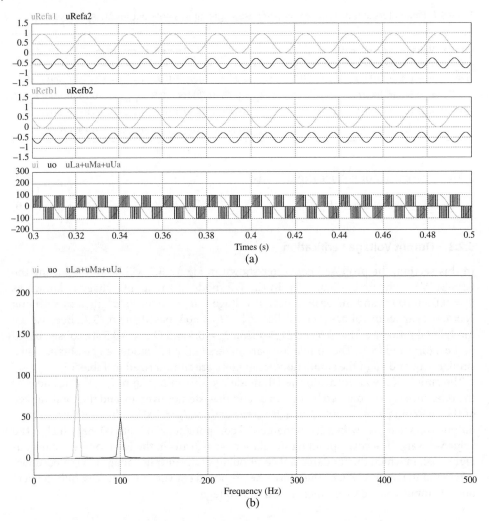

Figure 5.3 Simulation waveforms of the proposed single-phase single-input single-output AC–AC converter in the DF mode. (a) Time-domain waveforms. (b) Spectrum of u_i, u_o and total switching arm voltage.

5.3.2 Control Scheme

When the CPS-SPWM scheme is used to control the proposed converter, the total voltage of the switching arms in each phase will be NU_C. Similar to Eqs. (5.3) and (5.4), the reference signals for controlling the switching arm voltages can be defined by

$$u_{\text{Refx1}} = \frac{u_{\text{Lx}} + u_{\text{Mx}} - u_{\text{Ux}}}{NU_C} = \frac{2u_{\text{ix}}}{NU_C} + O_{\text{si}} = M_i \sin(2\pi f_i t + \varphi_i + \phi_x) + O_{\text{si}} \quad (5.12)$$

$$u_{\text{Refx2}} = \frac{u_{\text{Lx}} - u_{\text{Mx}} - u_{\text{Ux}}}{NU_C} = \frac{2u_{\text{ox}}}{NU_C} + O_{\text{so}} = M_o \sin(2\pi f_o t + \varphi_o + \phi_x) + O_{\text{so}} \quad (5.13)$$

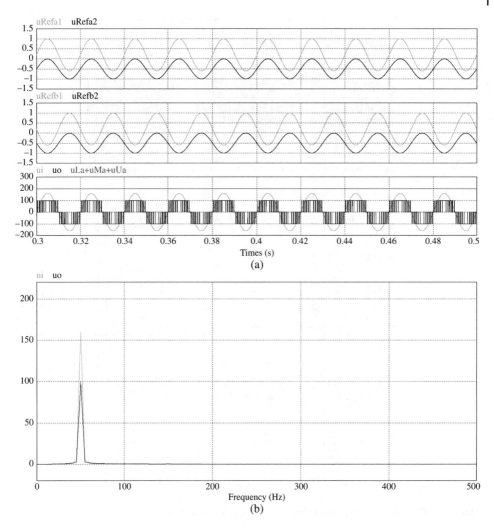

Figure 5.4 Simulation waveforms of the proposed single-phase single-input single-output AC–AC converter in the EF′ mode. (a) Time-domain waveforms. (b) Spectrum of u_i, u_o and total switching arm voltage.

where $x = $ u, v, w and $\phi_u = 0$, $\phi_v = -\frac{2\pi}{3}$, $\phi_w = \frac{2\pi}{3}$. $M_i = \frac{2U_i}{NU_C}$ and $M_o = \frac{2U_o}{NU_C}$ are modulated ratios, while U_i and U_o are phase voltage amplitudes, f_i and f_o are frequencies, ϕ_i and ϕ_o are phase shifts of the input AC voltage source and output AC voltage, respectively. O_{si} and O_{so} are DC offsets of the reference signals.

The relationship between modulated ratios and DC offsets is consistent with that in Section 5.2.2, that is,

$$\begin{cases} M_i + M_o \leq O_{si} - O_{so} & \text{except for EF}' \text{ mode} \\ |M_i - M_o| \leq O_{si} - O_{so} & \text{only for EF}' \text{ mode} \end{cases} \tag{5.14}$$

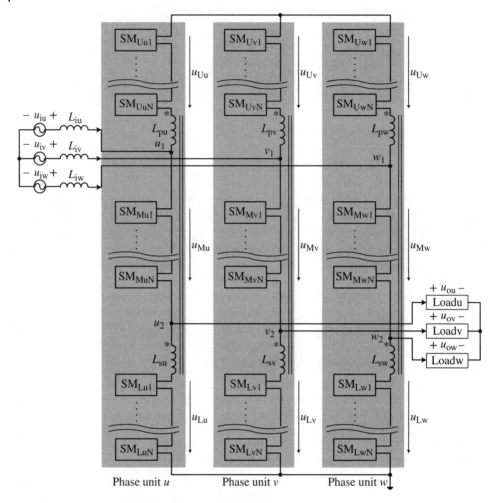

Figure 5.5 Three-phase single-input single-output AC–AC converter.

Considering the definition and range of M_i and M_o, the capacitor voltage of the SM should satisfy the following equation:

$$U_C \geq \begin{cases} \frac{2}{N}(U_i + U_o) & \text{except for EF}' \text{ mode} \\ \frac{2}{N} \times \max(U_i, U_o) & \text{only for EF}' \text{ mode} \end{cases} \tag{5.15}$$

5.3.3 Output Voltage Verification

In this section, the proposed converter in the DF mode will be verified by simulation results. If the voltage amplitudes of the known input voltage and the desired output voltage are $U_i = 50\,\text{V}$ and $U_o = 30\,\text{V}$, respectively, then the capacitor voltage could be selected as $U_C = 50\,\text{V}$ with $N = 4$ based on Eq. (5.15). The parameters of the reference signals can be obtained from Eqs. (5.12) to (5.14), as listed in Table 5.2.

When the output load of each phase is set to $L_{AC} = 0.1\,\text{H}$ and $R_{AC} = 10\,\Omega$, the output voltage and current of each phase in Case 1 are shown in Figure 5.6; it is found that the

Table 5.2 Parameters of the proposed three-phase single-input single-output AC–AC converter.

Case	u_{Ref1}				u_{Ref2}			
	M_i	f_i (Hz)	φ_i	O_{si} (V)	M_o	f_o (Hz)	φ_o	O_{so} (V)
1	0.5	50	0	0.4	0.3	100	0	−0.6
2	0.5	50	0	0.4	0.3	20	0	−0.6

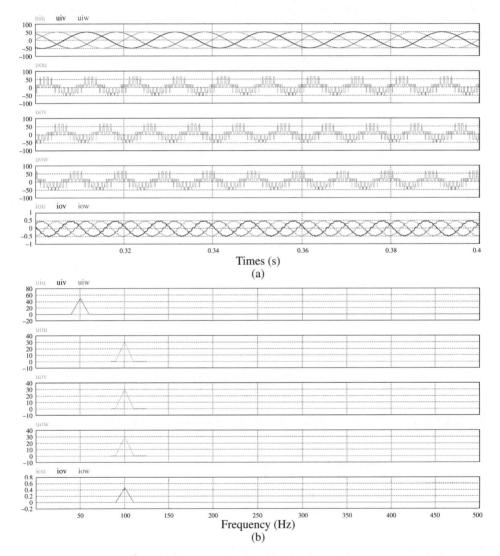

Figure 5.6 Simulation waveforms of the proposed three-phase single-input single-output AC–AC converter in Case 1. (a) Time-domain waveforms. (b) Spectrums.

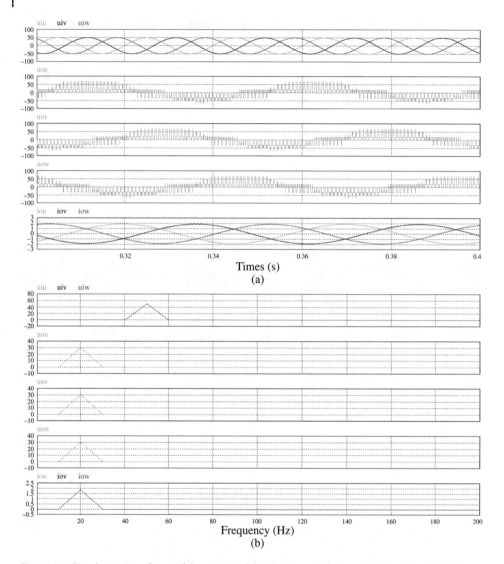

Figure 5.7 Simulation waveforms of the proposed three-phase single-input single-output AC–AC converter in Case 2. (a) Time-domain waveforms. (b) Spectrums.

fundamental frequency and the magnitude of the output voltage are 100 Hz and 30 V as desired. By changing the frequency of u_{Ref2} to 20 Hz, the corresponding simulation waveforms are provided in Figure 5.7. It is obvious that the three-phase input voltage can be converted directly to another three-phase voltage with higher or lower frequency.

5.4 Single-Phase Multiple-terminal AC–AC Converter

5.4.1 Basic Structure and Operating Principle

If there are more than one inputs or outputs, then the single-phase multiple-terminal AC–AC converter can be obtained by inserting some switching arms, as illustrated in Figure 5.8 [3]. The branch between two general phase units could be the AC input

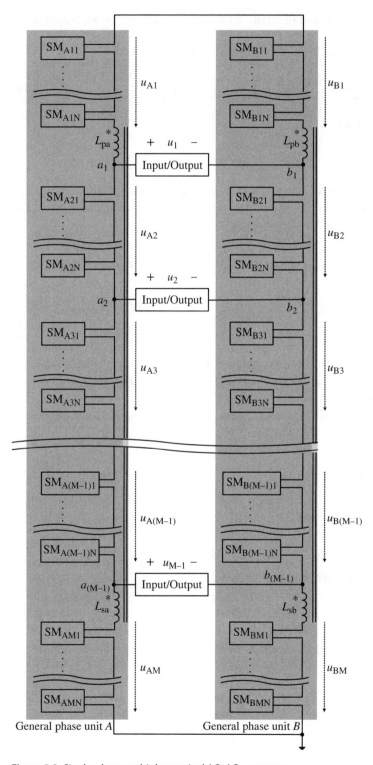

Figure 5.8 Single-phase multiple-terminal AC–AC converter.

source or the AC output load, but there should be at least one input or one output in the proposed converter.

Similar to Eq. (4.48), the voltage of the kth branch is determined by

$$u_{\text{akbk}} = \frac{1}{2}\left(-\sum_{i=1}^{k} u_{\text{Ai}} + \sum_{j=k+1}^{M} u_{\text{Aj}} + \sum_{i=1}^{k} u_{\text{Bi}} - \sum_{j=k+1}^{M} u_{\text{Bj}}\right) \tag{5.16}$$

where $k = 1, 2, \ldots, M - 1$.

5.4.2 Control Scheme

By applying the CPS-SPWM scheme to the proposed converter, the total of the switching-arm voltages in each general phase unit is controlled to be a stable DC voltage NU_C, then the reference signals are defined by

$$\begin{cases} u_{\text{Refak}} = \dfrac{1}{NU_C}\left(-\sum_{i=1}^{k} u_{\text{Ai}} + \sum_{j=k+1}^{M} u_{\text{Aj}}\right) = \dfrac{u_k}{NU_C} + O_{\text{sk}} = M_k \sin(2\pi f_k t + \phi_k) + O_{\text{sk}} \\[4mm] u_{\text{Refbk}} = \dfrac{1}{NU_C}\left(-\sum_{i=1}^{k} u_{\text{Bi}} + \sum_{j=k+1}^{M} u_{\text{Bj}}\right) = -\dfrac{u_k}{NU_C} + O_{\text{sk}} = -M_k \sin(2\pi f_k t + \phi_k) + O_{\text{sk}} \end{cases}$$

$$\tag{5.17}$$

where $M_k = \dfrac{U_k}{NU_C}$ and O_{sk} are the modulated ratio and the DC offset of the kth reference signal, respectively. U_k, f_k, and φ_k are amplitude, frequency, and phase shift of the kth branch voltage, respectively.

Referring to the analysis in Section 4.6.2, the DC offsets should satisfy the following equation in the case of DF and EF modes:

$$\begin{cases} O_{\text{s1}} \leq 1 - M_1 \\ O_{\text{s2}} \leq O_{\text{s1}} - M_1 - M_2 \\ \cdots \\ O_{\text{sk}} \leq O_{\text{s(k-1)}} - M_{k-1} - M_k \\ \cdots \\ O_{\text{s(M-1)}} \leq O_{\text{s(M-2)}} - M_{M-2} - M_{M-1} \leq 1 - M_{M-1} \end{cases} \tag{5.18}$$

and the corresponding capacitor voltage should satisfy

$$U_C \geq \frac{1}{N} \sum_{k=1}^{M-1} U_k \tag{5.19}$$

In the case of EF′ mode, the DC offsets of the nearby references should meet the requirement

$$O_{\text{s(k-1)}} - O_{\text{sk}} \geq |M_k - M_{k-1}| \tag{5.20}$$

and the capacitor voltage must be larger than the value defined by the following equation:

$$U_C > \frac{1}{N} \times \max(U_1, U_2, \ldots, U_{M-1}) \tag{5.21}$$

Table 5.3 Parameters of the proposed single-phase multiple-terminal AC–AC converter.

Mode	Figure	Terminal	Voltage expression (V)	M_k	O_{sk} (V)	f_k (Hz)	φ_k(rad)
EF′	5.9	Input 1	$u_1 = 160 \sin 100\pi t$	0.4	0.6	50	0
		Input 2	$u_2 = 120 \sin 100\pi t$	0.3	0.2	50	0
		Output 1	$u_3 = 240 \sin 100\pi t$	0.6	−0.4	50	0
	5.10	Input	$u_1 = 300 \sin 100\pi t$	0.75	0.2	50	0
		Output	$u_2 = 150 \sin 100\pi t$	0.375	−0.3	50	0
		Output	$u_3 = 100 \sin 100\pi t$	0.25	−0.7	50	0
DF	5.11	Input	$u_1 = 100 \sin 100\pi t$	0.25	0.7	50	0
		Input	$u_2 = -100 \sin 100\pi t$	0.25	0.0	50	π
		Output	$u_3 = 110 \sin 120\pi t$	0.275	−0.7	60	0
	5.12	Input	$u_1 = 220 \sin 100\pi t$	0.55	0.4	50	0
		Output	$u_2 = 110 \sin 120\pi t$	0.275	−0.45	60	0
		Output	$u_3 = 50 \sin 100\pi t$	0.125	−0.85	50	0

5.4.3 Output Voltage Verification

A single-phase three-terminal AC–AC prototype, that is M = 4, is built to verify the performance of the proposed converter. As there are three terminals, there are two kinds of converter configuration, one dual-input single-output and the other single-input dual-output. If N = 4 and $U_C = 100$ V are selected, then the simulation parameters of different modes are as listed in Table 5.3, derived from Eqs. (5.18)–(5.21).

The simulation waveforms of the reference voltages, input and output voltages under four different situations are shown in Figures 5.9–5.12, respectively. It is obvious that the fundamental component of the output voltage agrees with the desired expression in each case. Therefore, the proposed single-phase multiple-terminal AC–AC converter has been verified.

5.5 Three-Phase Multiple-terminal AC–AC Converter

5.5.1 Basic Structure and Operating Principle

It is obvious that the three-phase multiple-terminal AC–AC converter can be obtained just by adding one general phase unit to Figure 5.8 [4]. As shown in Figure 5.13, if there are M switching arms in the general phase unit, then there are M − 1 input/output ports in the proposed converter, and the minimum number of inputs/outputs should be one.

Similar to Section 4.7.1, the voltage between the switching arms at steady state can be expressed by

$$u_{kx} - u_{ky} = \frac{1}{2}\left(-\sum_{i=1}^{k} u_{Xi} + \sum_{j=k+1}^{M} u_{Xj} + \sum_{i=1}^{k} u_{Yi} - \sum_{j=k+1}^{M} u_{Yj} \right) \tag{5.22}$$

where $k = 1, 2, \ldots, M - 1$; $x, y = \{u, v, w\}$, and X, Y = $\{U, V, W\}$.

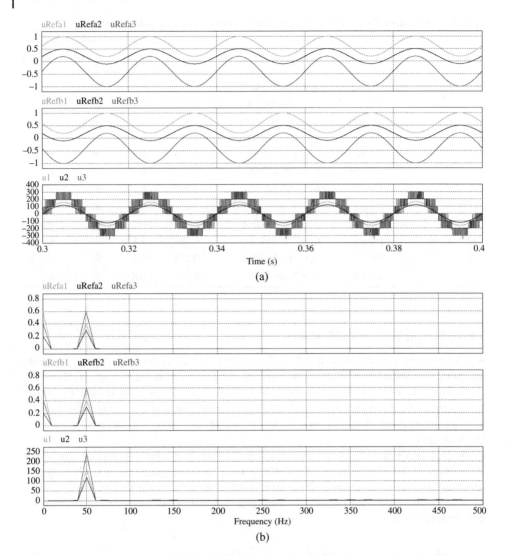

Figure 5.9 Simulation waveforms of the proposed single-phase dual-input single-output AC–AC converter in the EF′ mode. (a) Time-domain waveforms. (b) Spectrums.

5.5.2 Control Scheme

Normally, the three-phase input/output voltage has the unified form

$$u_{kx}(t) = U_k \sin(2\pi f_k t + \varphi_k + \phi_x) \tag{5.23}$$

where $\phi_u = 0$, $\phi_v = -\frac{2\pi}{3}$, $\phi_w = \frac{2\pi}{3}$, U_k, f_k, and ϕ_k are the amplitude, frequency, and phase shift of the AC voltage, respectively.

If the CPS-SPWM scheme is selected to control the proposed three-phase AC–AC converter, and the capacitor voltage U_C is kept constant, then the reference signals to

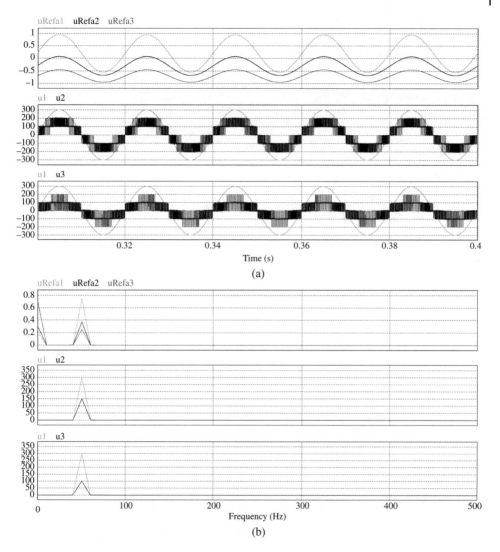

Figure 5.10 Simulation waveforms of the proposed single-phase single-input dual-output AC–AC converter in the EF′ mode. (a) Time-domain waveforms. (b) Spectrums.

regulate the switching-arm voltages should be determined based on Eq. (5.22), that is

$$
\begin{cases}
u_{\text{Refkx}} = \dfrac{1}{NU_C}\left(-\sum_{i=1}^{k} u_{Xi} + \sum_{j=k+1}^{M} u_{Xj}\right) = \dfrac{2u_{kx}}{NU_C} + U_{\text{osk}} \\
\qquad = M_k \sin(2\pi f_k t + \phi_k + \varphi_x) + O_{\text{sk}} \\
u_{\text{Refky}} = \dfrac{1}{NU_C}\left(-\sum_{i=1}^{k} u_{Yi} + \sum_{j=k+1}^{M} u_{Yj}\right) = \dfrac{2u_{ky}}{NU_C} + U_{\text{osk}} \\
\qquad = M_k \sin(2\pi f_k t + \phi_k + \varphi_y) + O_{\text{sk}}
\end{cases}
\tag{5.24}
$$

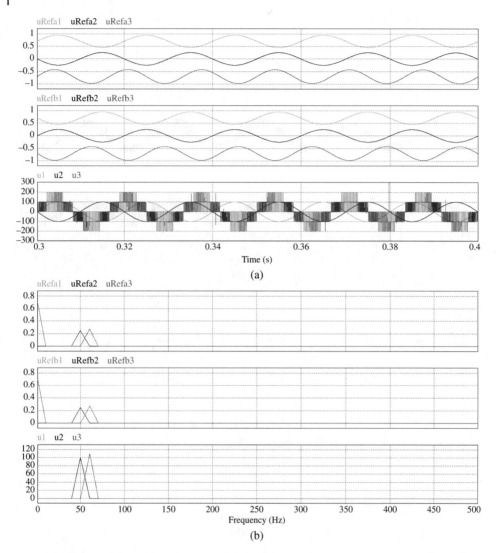

Figure 5.11 Simulation waveforms of the proposed single-phase dual-input single-output AC–AC converter in the DF mode. (a) Time-domain waveforms. (b) Spectrums.

where $M_k = \frac{2U_k}{NU_C}$ and O_{sk} are the modulated ratio and the DC offset of the kth reference signals, respectively.

The requirement of the DC offsets in different cases are the same as in Eqs. (5.18) and (5.20). Accordingly, the capacitor voltage should be

$$U_C \geq \frac{2}{N} \sum_{k=1}^{M-1} U_k \qquad \text{for the DF and EF modes} \qquad (5.25)$$

$$U_C > \frac{2}{N} \times \max(U_1, \ U_2, \dots, \ U_{M-1}) \quad \text{for the EF' mode} \qquad (5.26)$$

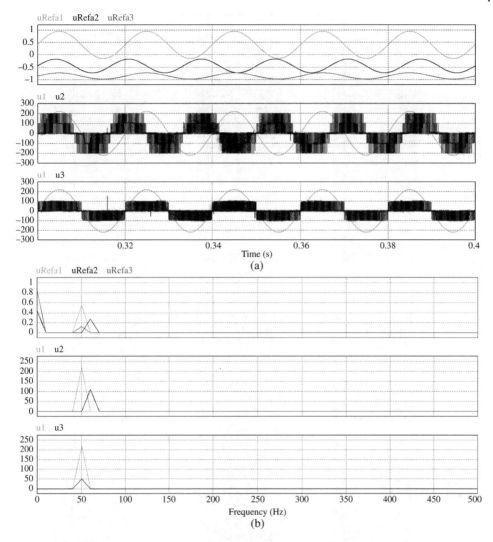

Figure 5.12 Simulation waveforms of the proposed single-phase single-input dual-output AC–AC converter in the DF mode. (a) Time-domain waveforms. (b) Spectrums.

5.5.3 Output Voltage Verification

In this section, a prototype with M = 4 is taken as an example to verify the performance of the proposed multiple-terminal AC–AC converters. The simulation parameters when the prototype is designed to operate in the DF mode, which is the more popular case, are listed in Table 5.4. Based on the known input voltage and the desired output voltage, $U_C = 100$ V when N = 4, and the reference signals are defined by Eqs. (5.18) and (5.24).

When both u_{1x} and u_{2x} are input voltage sources, the simulation waveforms and corresponding spectrums of reference signals in phase unit u, the terminal voltages in phase

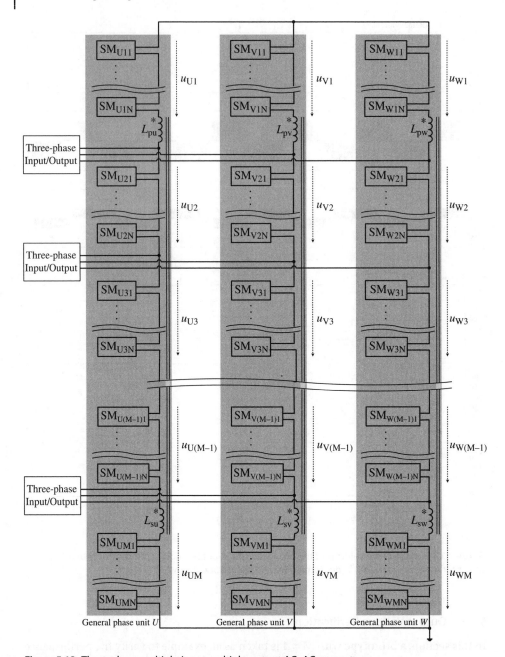

Figure 5.13 Three-phase multiple-input multiple-output AC–AC converter.

unit u, one of the output line voltages, and the total voltage of the phase unit (UDC) are given in Figure 5.14. Though there are some noises in the output phase and line voltages, the fundamental component of the output phase voltage u_{3u} is the same as the desired value.

Table 5.4 Parameters of the proposed three-phase multiple-terminal AC–AC converter.

Case	Terminal k	Voltage expression u_{kx} (V)	M_k	O_{sk} (V)	f_k (Hz)	φ_k (rad)
1	1 (input)	$u_{1x} = 40\sin(100\pi t + \varphi_x)$	0.2	0.8	50	0
	2 (input)	$u_{2x} = 60\sin(200\pi t + \varphi_x)$	0.3	0.3	100	0
	3 (output)	$u_{3x} = 100\sin(120\pi t + \varphi_x)$	0.5	−0.5	60	0
2	1 (input)	$u_{1x} = 110\sin(120\pi t + \varphi_x)$	0.55	0.45	60	0
	2 (output)	$u_{2x} = 40\sin(100\pi t + \varphi_x)$	0.2	−0.35	50	0
	3 (output)	$u_{3x} = 40\cos(100\pi t + \varphi_x)$	0.2	−0.8	50	$\pi/2$

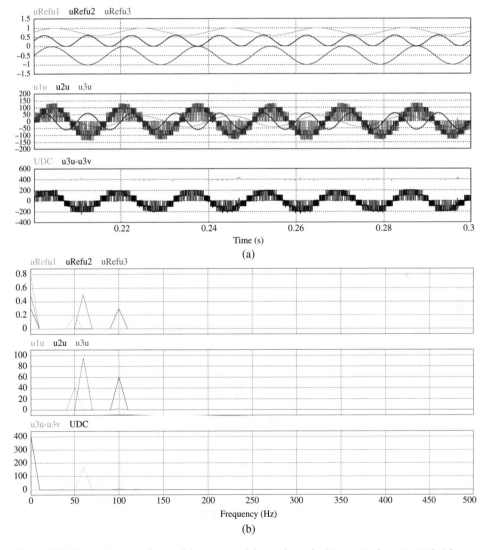

Figure 5.14 Simulation waveforms of the proposed three-phase dual-input single-output AC–AC converter. (a) Time-domain waveforms. (b) Spectrums.

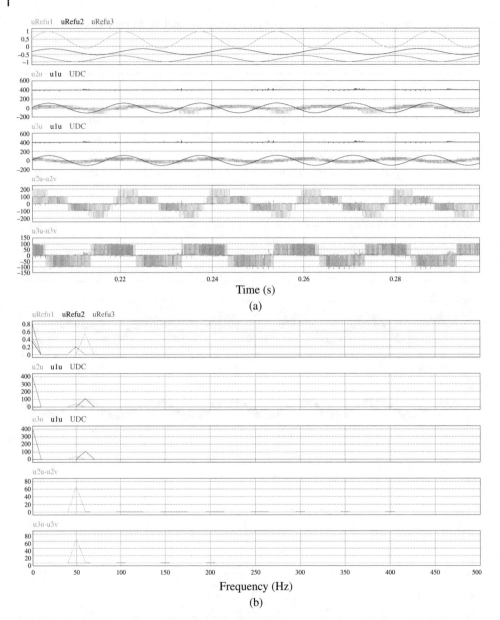

Figure 5.15 Simulation waveforms of the proposed three-phase single-input dual-output AC–AC converter. (a) Time-domain waveforms. (b) Spectrums.

The simulation results when only u_{1x} is the input voltage source are shown in Figure 5.15, in which both output phase voltages u_{2u} and u_{3u} have the designed fundamental components. It is concluded that the proposed three-phase multiple-terminal AC–AC converter can operate in the multiple-input multiple-output situation.

5.6 Summary

In this chapter, multiple-input multiple-output high-voltage AC–AC converters are proposed for the first time, which can realize power transfer from more than one AC source to more than one AC output directly. The unique characteristics of the proposed converter include that no high-voltage DC link capacitor is needed and only one converter (which has fewer components) is used, which will be suitable for the multi-terminal high-voltage AC systems.

References

1 Zhang, B., Fu, J., Qiu, D. Y. Single-phase six switching-groups MMC AC-AC converter and its control method. State Intellectual Property Office of the P.R.C., ZL 2014101193815, 2016.9.21.

2 Zhang, B., Fu, J., Qiu, D. Y. Three-phase nine switching-groups MMC AC-AC converter and its control method. State Intellectual Property Office of the P.R.C., ZL 201410119191.3, 2016.10.5.

3 Zhang, B., Fu, J., Qiu, D. Y. 2N+2 switching-groups MMC AC-AC converter and its control method. State Intellectual Property Office of the P.R.C., ZL 201410385037.0, 2017.10.20.

4 Zhang, B., Fu, J., Qiu, D. Y. 3N+3 switching-groups MMC AC-AC converter. State Intellectual Property Office of the P.R.C., ZL 201420144429.3, 2014.9.10.

6

Multiple-terminal High-voltage DC–DC Converters

6.1 Introduction

In many power electronic applications ranging from renewable energy systems to electric vehicles, and laptop power supplies, energy transfer between multiple sources, loads, and energy storage components may be required. In order to effectively adapt to this energy system architecture, multi-terminal DC–DC converter topologies provide a more cost-effective solution [1]. Based on the operating principle of the proposed multi-terminal high-voltage converters in Chapters 3–5, the voltage of the switching arm is a combination of several capacitor voltages, and the voltage on the endpoint of the switching arm has been proven to be of DC value [2]. Therefore, taking the phase unit or general phase unit consisting of several switching arms as a single converter, a multiple-output high-voltage DC–DC converter with single or multiple inputs can be constructed.

6.2 Single-Input Dual-Output DC–DC Converter

6.2.1 Basic Structure and Operating Principle

By taking the phase unit defined in the aforementioned multi-terminal high-voltage converters as a single converter, the proposed single-input dual-output DC–DC converter is illustrated in Figure 6.1, consisting of three switching arms and one coupled inductor. The endpoints of the switching arm can be considered as the input/output terminals, then there are three terminals T_1, T_2, and T_3 in the proposed converter.

Referring to the voltage analysis in Section 3.2, the relationships between three terminal voltages, u_1, u_2, and u_3, are expressed by

$$\begin{cases} u_1 = u_2 + u_3 + u_{A1} - u_{A3} \\ u_2 = \frac{1}{2}(u_1 + u_{A3} + u_{A2} - u_{A1}) \\ u_3 = \frac{1}{2}(u_1 + u_{A3} - u_{A2} - u_{A1}) \end{cases} \tag{6.1}$$

where u_{A1}, u_{A2}, and u_{A3} are the output voltages of the switching arm, respectively.

If there are N sub-modules (SMs) in each switching arm and the capacitor voltage of the SM is kept at a constant value U_C, then the output voltage of the switching arm, u_{A1}, u_{A2}, or u_{A3}, varies in the range from 0 to NU_C, and we have

$$U_1 > U_2 > U_3 \tag{6.2}$$

where U_1, U_2, and U_3 are the average values of u_1, u_2, and u_3, respectively.

Multi-terminal High-voltage Converter, First Edition. Bo Zhang and Dongyuan Qiu.
© 2019 John Wiley & Sons Singapore Pte. Ltd. Published 2019 by John Wiley & Sons Singapore Pte. Ltd.

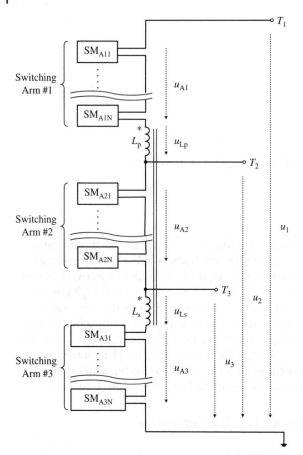

Figure 6.1 Single-input dual-output DC–DC converter.

If one and only one of the three terminals is connected to the DC voltage source, then the proposed DC–DC converter has three kinds of single-input configuration, as shown in Figure 6.2.

6.2.2 Control Scheme

As both the input and output of the proposed single-input dual-output converter are DC voltages, if the carrier phase-shifted sinusoidal pulse-width modulation (CPS-SPWM) scheme introduced in Chapters 3–5 is used, then the reference signals should be DC ones, as shown in Figure 6.3a. Similar to Figure 3.4, the method to generate the control signals for the SMs, which is renamed as the carrier phase-shifted pulse-width modulation (CPS-PWM) scheme, is illustrated in Figure 6.3b.

Thus, based on the operating principle of the control scheme, the total output voltage of the three switching arms will be $u_{A1} + u_{A2} + u_{A3} = NU_C$, and the definition of the reference signals can be expressed by

$$
\begin{cases}
u_{Ref1} = \dfrac{u_{A3} + u_{A2} - u_{A1}}{NU_C} \\[2mm]
u_{Ref2} = \dfrac{u_{A3} - u_{A2} - u_{A1}}{NU_C}
\end{cases}
\tag{6.3}
$$

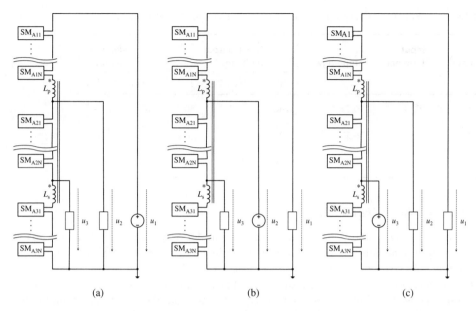

Figure 6.2 Different configurations of the single-input dual-output DC–DC converter. (a) T_1 is the input terminal. (b) T_2 is the input terminal. (c) T_3 is the input terminal.

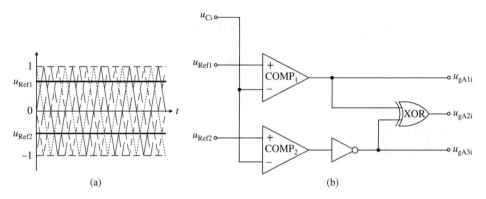

Figure 6.3 Control schematic of the single-input dual-output DC–DC converter. (a) Typical reference signals. (b) CPS-PWM scheme.

Based on Eq. (6.1), Eq. (6.3) turns out to be

$$\begin{cases} u_{\text{Ref1}} = \dfrac{u_{A3} + u_{A2} - u_{A1}}{NU_C} = \dfrac{2U_2 - U_1}{NU_C} \\[4mm] u_{\text{Ref2}} = \dfrac{u_{A3} - u_{A2} - u_{A1}}{NU_C} = \dfrac{2U_3 - U_1}{NU_C} \end{cases} \tag{6.4}$$

It is obvious that $1 > u_{\text{Ref1}} > u_{\text{Ref2}} > -1$, which means that there will be no overlap for the output voltages.

Table 6.1 Parameters of the proposed single-input dual-output DC–DC converter.

Case	Input terminal	Topology	Input/output voltages (V)			Reference signals	
			U_1	U_2	U_3	u_{Ref1}	u_{Ref2}
1	T_1	6.2a	240	120	80	0	$-\dfrac{1}{3}$
2	T_2	6.2b	240	150	80	$\dfrac{1}{4}$	$-\dfrac{1}{3}$
3	T_3	6.2c	240	120	100	0	$-\dfrac{1}{6}$

6.2.3 Simulation Verification

When N = 4, $U_C = 60$ V, different groups of simulation parameters of the proposed single-input dual-output DC–DC converter are listed in Table 6.1, in which the average terminal voltages should satisfy $NU_C \geq U_1 > U_2 > U_3$. The simulation waveforms of the three different configurations are shown in Figure 6.4; some of the output voltages are pulsed voltage, but their average value is equal to the desired one. Therefore, no matter which configuration the proposed DC–DC converter has, the desired output voltages can be obtained by controlling the switching-arm voltages, u_{A1}, u_{A2}, and u_{A3}, based on Eq. (6.4).

6.3 Single-Input Multiple-Output DC–DC Converter

6.3.1 Basic Structure and Operating Principle

If the phase unit in Figure 6.1 is replaced by the general phase unit with multiple switching arms, then the proposed single-input multiple-output DC–DC converter is illustrated in Figure 6.5. As shown in Figure 6.5, there are M terminals if the number of switching arms is M.

According to the voltage analysis in Section 3.5.1, the terminal voltages are determined by

$$\begin{cases} u_1 = \sum_{i=1}^{M} u_{Ai} + u_{Lp} + u_{Ls} \\ u_2 = u_1 - u_{A1} - u_{Lp} \\ u_3 = u_2 - u_{A2} \\ \cdots \\ u_{x+1} = u_x - u_{Ax} \\ \cdots \\ u_M = u_{M-1} - u_{A(M-1)} = u_{AM} + u_{Ls} \end{cases} \qquad (6.5)$$

where u_{Ax} represents the output voltage of the xth switching arm, x = 1, 2, ..., M. Assume that $u_{Lp} = u_{Ls}$, then we have

$$u_1 = u_2 + u_M + u_{A1} - u_{AM} \qquad (6.6)$$

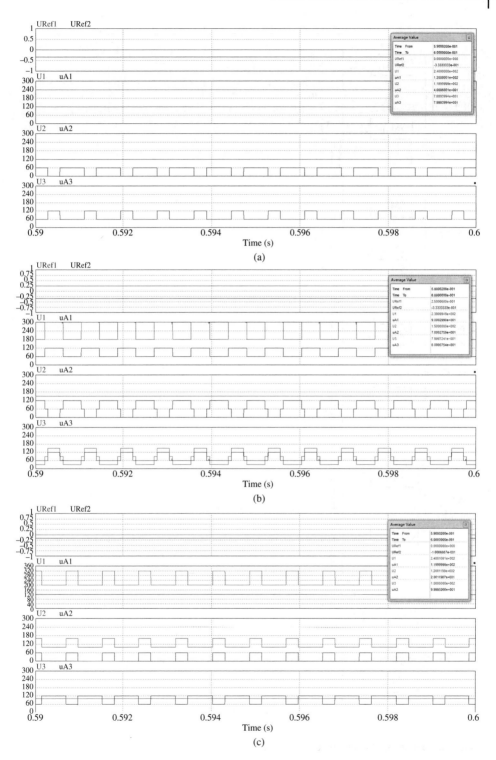

Figure 6.4 Simulation waveforms of the proposed single-input dual-output DC–DC converter. (a) Case 1, T_1 is the input terminal. (b) Case 2, T_2 is the input terminal. (c) Case 3, T_3 is the input terminal.

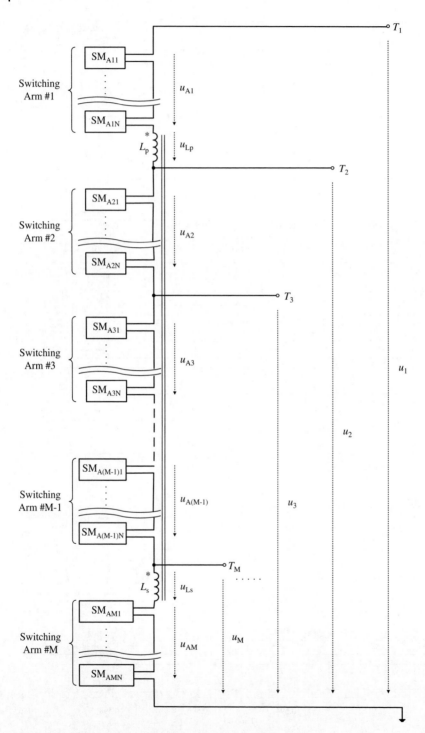

Figure 6.5 Single-input multiple-output DC–DC converter.

The general form of the terminal voltage, except for u_1, is

$$u_k = \frac{1}{2}\left(u_1 - \sum_{i=2}^{k-1} u_{Ai} + \sum_{j=k}^{M} u_{Aj} \right) \tag{6.7}$$

where $2 \leq k \leq M$.

The average value of the terminal voltage satisfies the following equation:

$$U_1 > \cdots > U_k > U_{k+1} > \cdots > U_M \tag{6.8}$$

As there is only one terminal connected to the DC voltage source, the proposed single-input multiple-output DC–DC converter will have M kinds of single-input configuration. Among them, the case when T_1 is the input terminal can be considered as the buck mode because all of the output voltages are lower than the input one, while the case when T_M is the input terminal can be considered as the boost mode.

6.3.2 Control Scheme

Based on Section 3.5.2, the method to generate the control signals for the SMs is illustrated in Figure 6.6, when the CPS-PWM scheme is applied to the proposed single-input multiple-output DC–DC converter.

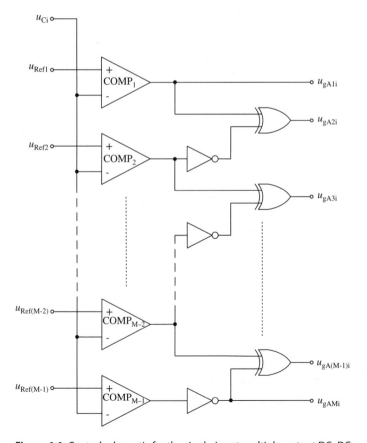

Figure 6.6 Control schematic for the single-input multiple-output DC–DC converter.

Table 6.2 Parameters of the proposed single-input multiple-output DC–DC converter.

Case	Input terminal	Input/output voltages (V)				Reference signals		
		U_1	U_2	U_3	U_4	u_{Ref1}	u_{Ref2}	u_{Ref3}
1	T_1	240	150	120	48	0.25	0	−0.6
2	T_4	200	160	100	48	0.5	0	−13/30

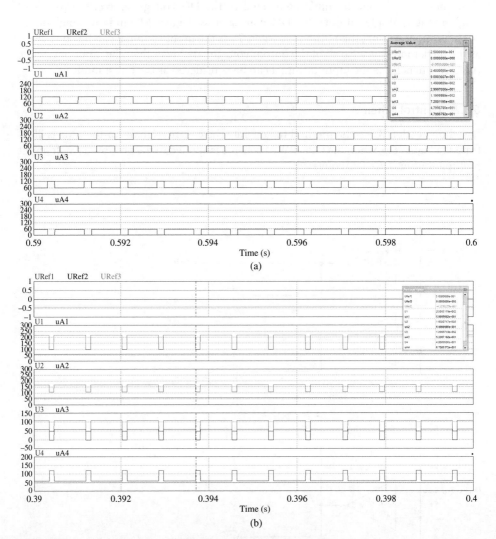

Figure 6.7 Simulation waveforms of the proposed single-input multiple-output DC–DC converter. (a) Case 1, T_1 is the input terminal. (b) Case 2, T_4 is the input terminal.

If there are N SMs in each switching arm and the capacitor voltage of the SM remains at U_C, then the total output voltage of all switching arms will be

$$\sum_{x=1}^{M} u_{Ax} = NU_C \tag{6.9}$$

Thus, the definition of the reference signal can be expressed by

$$u_{Ref(k-1)} = \frac{\sum_{i=2}^{k-1} u_{Ai} - \sum_{j=k}^{M} u_{Aj}}{\sum_{x=1}^{M} u_{Ax}} = \frac{2u_k - u_1}{NU_C} \tag{6.10}$$

where $2 \leq k \leq M$.

If the input voltage and all of the desired output voltages are known, then Eq. (6.10) is simplified to

$$u_{Refk} = \frac{2U_k - U_1}{NU_C} \tag{6.11}$$

Based on Eqs. (6.8) and (6.9), it is known that $NU_C \geq U_1 > \cdots > U_k > U_{k+1} > \cdots > U_M$, then $1 > u_{Ref1} > u_{Ref2} > \cdots > u_{Ref(M-1)} > -1$, which results in no overlaps of the output voltages.

6.3.3 Simulation Verification

A prototype with four switching arms (M = 4) or three output terminals is used as an example to verify the performance of the proposed single-input multiple-output DC–DC converter. Select N = 4, U_C = 60 V; the other simulation parameters are listed in Table 6.2, and the corresponding simulation waveforms are shown in Figure 6.7.

From the simulation results, it is found that the output voltages are very well consistent with the theoretical analysis.

6.4 Multiple-Input Multiple-Output DC–DC Converter

From Eq. (6.7), it is found that the voltage difference between any two terminals, except for T_1 and T_M, is only determined by the switching arm voltages

$$u_x - u_y = \sum_{i=x}^{y-1} u_{Ai} \tag{6.12}$$

where $1 < x < y < M$.

As the output voltage of the SM is 0 or U_C, the instantaneous voltage of the switching arm will be 0, U_C, $2U_C$, ..., or NU_C. If the result of Eq. (6.12) is equal to several times U_C, then these two terminal voltages will be constant, which means that they are able to connect with the input DC voltage sources.

Therefore, the converter in Figure 6.5 can be used as a multiple-input multiple-output DC–DC converter if and only if the input voltages are connected to any terminals, except for T_1 and T_M, and meet the requirement of

$$U_x - U_y = \lambda U_C \tag{6.13}$$

where λ is an integer.

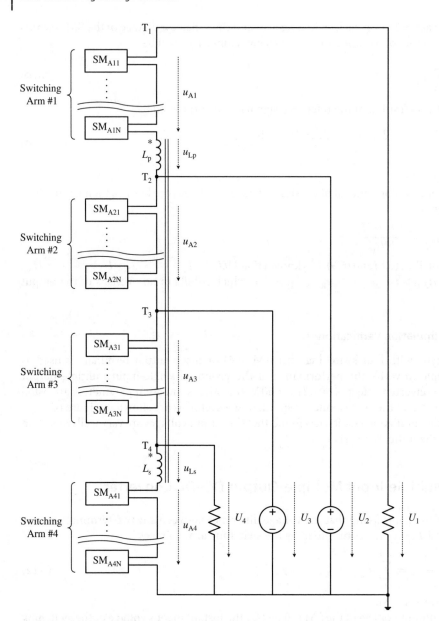

Figure 6.8 A dual-input dual-output DC–DC converter.

Table 6.3 Parameters of the proposed dual-input dual-output DC–DC converter.

Case	Input voltages (V)		Output voltages (V)		Reference signals (V)		
	U_2	U_3	U_1	U_4	u_{Ref1}	u_{Ref2}	u_{Ref3}
1	150	100	180	48	0.6	0.1	−0.42
2	170	120	200	40	0.5	0	−0.45
3	175	75	200	36	0.75	−0.25	−0.64

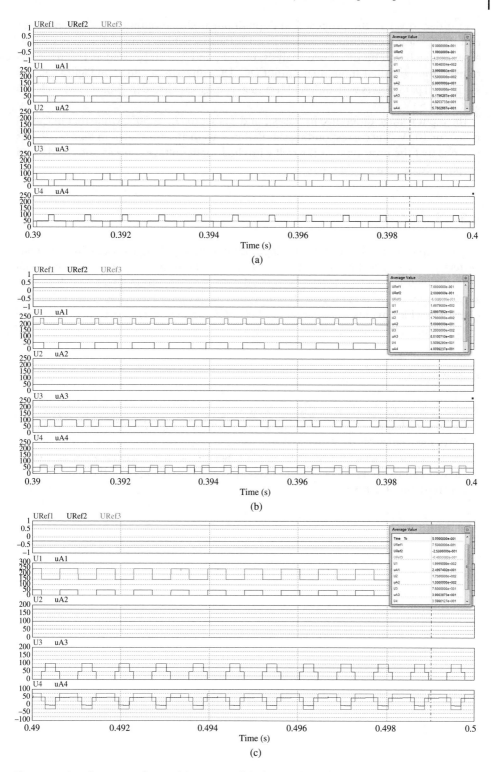

Figure 6.9 Simulation waveforms of the proposed dual-input dual-output DC–DC converter.
(a) Case 1. (b) Case 2. (c) Case 3.

In order to verify the proposed multiple-input multiple-output DC–DC converter and its operating condition, a topology with four switching arms (M = 4) is used as an example. Based on the above analysis, only T_2 and T_3 can be used as the input terminals. Thus, the dual-input dual-output DC–DC converter is shown in Figure 6.8. The simulation parameters with different terminal voltages are listed in Table 6.3 when N = 4 and $U_C = 50$ V.

The input voltage difference in Cases 1 and 2 is set to U_C or 50 V, while that in Case 3 is $2U_C$ or 100 V. The corresponding simulation waveforms are shown in Figure 6.9. It is obvious that the prototype works well with dual-input voltage sources and the desired output voltages can be obtained.

6.5 Summary

In this chapter, multiple-output with single or multiple-input high-voltage DC–DC converters are proposed for the first time, which can be used in applications requiring different high DC voltage levels. Compared with the conventional way, which uses N single-output converters for N voltage outputs, the proposed multiple-output DC–DC converter, which avoids using large numbers of components, is a better way to realize the multiple-voltage supply.

References

1 Filsoof, K. and Lehn, P. W. (2016). A bidirectional multiple-input multiple-output modular multilevel DC-DC converter and its control design. *IEEE Transactions on Power Electronics* 31 (4): 2767–2779.
2 Zhang, B., Fu, J., Qiu, D. Y. Double-output single-phase three switching-groups MMC inverter and its control method. State Intellectual Property Office of the P.R.C., ZL 201410042986.9, 2016.6.22.

7

Multiple-terminal High-voltage Hybrid Converters

7.1 Introduction

Renewable energy sources, combined with energy storage systems and/or utility grids, can deliver continuous power to the loads. On such occasions, there are multiple energy sources with different types, for example, the outputs of solar array and fuel cell are of DC type, those of wind generator and utility are of AC type. Thus, it is predicted that a hybrid converter is able to interconnect various energy sources and loads with fewer converter stages and higher energy efficiency.

As illustrated in Chapters 3–6, the four basic kinds of multiple-terminal high-voltage power conversion, which are DC–AC, AC–DC, AC–AC, and DC–DC, can be obtained by controlling the switching arm voltages of the phase unit or general phase unit. It is possible to construct a multiple-terminal high-voltage hybrid converter in the similar way. Since DC distribution systems have attracted much attention recently, this chapter will focus on the configuration of the multiple-terminal hybrid converter with single DC output.

7.2 Six-Arm Hybrid Converter with Single-Phase AC Input

7.2.1 Basic Structure and Operating Principle

It is known from the multiple-input AC–DC converter proposed in Chapter 4 that several input AC sources can be connected between the switching arms, and it is also known from the multiple-terminal DC–DC converter proposed in Chapter 6 that the input DC source can be connected at the endpoint of the switching arm. In order to construct a hybrid converter with both AC and DC inputs, the single-phase six-arm dual-input single-output AC–DC converter shown in Figure 4.4 is used. If one of the AC sources, for example, u_2, is removed, and two DC voltage sources are connected to the endpoints a_2 and b_2, then the six-arm hybrid converter with single-phase AC input can be obtained, as demonstrated in Figure 7.1 [1].

Similar to the analysis in Section 4.3.1, the endpoint voltages of the switching arms are expressed by

$$
\begin{cases}
u_{a1} = \dfrac{U_O + u_{A3} + u_{A2} - u_{A1}}{2} \\
u_{a2} = \dfrac{U_O + u_{A3} - u_{A2} - u_{A1}}{2}
\end{cases}
\tag{7.1}
$$

Multi-terminal High-voltage Converter, First Edition. Bo Zhang and Dongyuan Qiu.
© 2019 John Wiley & Sons Singapore Pte. Ltd. Published 2019 by John Wiley & Sons Singapore Pte. Ltd.

$$\begin{cases} u_{b1} = \dfrac{U_O + u_{B3} + u_{B2} - u_{B1}}{2} \\[2mm] u_{b2} = \dfrac{U_O + u_{B3} - u_{B2} - u_{B1}}{2} \end{cases} \tag{7.2}$$

where u_{A1}, u_{A2}, u_{A3}, u_{B1}, u_{B2}, and u_{B3} are the switching-arm voltages.

As the single-phase voltage source u_{AC} is connected between terminals a_1 and b_1, the terminal voltages u_{a1} and u_{b1} should satisfy the following equation:

$$u_{a1} - u_{b1} = u_{ab1} = u_{AC} \tag{7.3}$$

Two DC voltage sources, u_{DCa} and u_{DCb}, are connected to terminals a_2 and b_2, respectively, then the terminal voltages u_{a2} and u_{b2} should be consistent with the following equations:

$$u_{a2} = U_{DCa} \tag{7.4}$$

$$u_{b2} = U_{DCb} \tag{7.5}$$

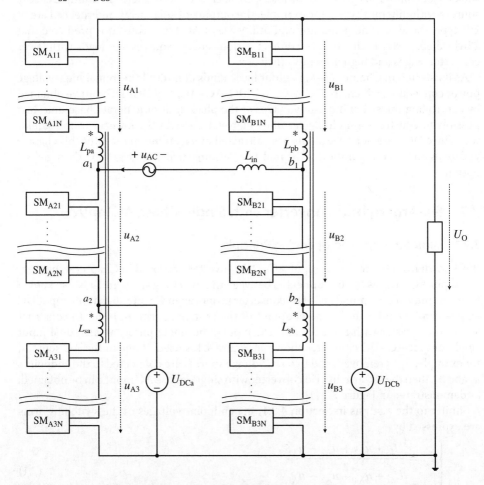

Figure 7.1 Six-arm hybrid converter with single-phase AC input.

Therefore, if Eqs. (7.3)–(7.5) are established by controlling the switching-arm voltages, the desired DC output voltage U_O can be obtained.

7.2.2 Control Scheme

Based on Eqs. (7.1)–(7.5), the relationship between the switching-arm voltages and the input voltages can be expressed by

$$(u_{A3} + u_{A2} - u_{A1}) - (u_{B3} + u_{B2} - u_{B1}) = 2u_{AC} \tag{7.6}$$

$$u_{A3} - u_{A2} - u_{A1} = 2U_{DCa} - U_O \tag{7.7}$$

$$u_{B3} - u_{B2} - u_{B1} = 2U_{DCb} - U_O \tag{7.8}$$

Assume that the single-phase input voltage is in the form $u_{AC} = U_{AC} \sin 2\pi ft$, where U_{AC} is the amplitude of the input voltage source. In order to obtain a constant output voltage U_O, the reference signals to control the switching-arm voltages can be defined by

$$u_{Refa1} = \frac{u_{A3} + u_{A2} - u_{A1}}{U_O} = \frac{u_{AC}}{U_O} + U_{os} = M \sin 2\pi ft + U_{os} \tag{7.9}$$

$$u_{Refb1} = \frac{u_{B3} + u_{B2} - u_{B1}}{U_O} = -\frac{u_{AC}}{U_O} + U_{os} = -M \sin 2\pi ft + U_{os} \tag{7.10}$$

$$u_{Refa2} = \frac{u_{A3} - u_{A2} - u_{A1}}{U_O} = \frac{2U_{DCa}}{U_O} - 1 \tag{7.11}$$

$$u_{Refb2} = \frac{u_{B3} - u_{B2} - u_{B1}}{U_O} = \frac{2U_{DCb}}{U_O} - 1 \tag{7.12}$$

where $M = \frac{U_{AC}}{U_O}$ is the modulated ratio and U_{os} is the DC offset to avoid overlap of the reference signals.

If the carrier phase-shifted sinusoidal pulse-width modulation (CPS-SPWM) scheme is used to control the switching-arm voltages, then the reference signals u_{Refa1} and u_{Refa2} are as illustrated in Figure 7.2.

Figure 7.2 Relationship between the carrier signals and the reference signals of the six-arm hybrid converter with single-phase AC input.

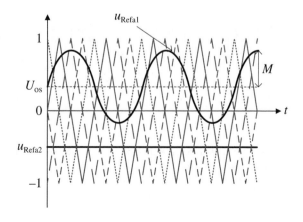

Table 7.1 Parameters of the six-arm hybrid converter with single-phase AC input.

Case	U_O (V)	U_{AC} (V)	f (Hz)	U_{DCa} (V)	U_{DCb} (V)	U_{os} (V)	Figure
1	200	100	50	50	50	0.2	7.3a
2	200	100	50	50	100	0.5	7.3b

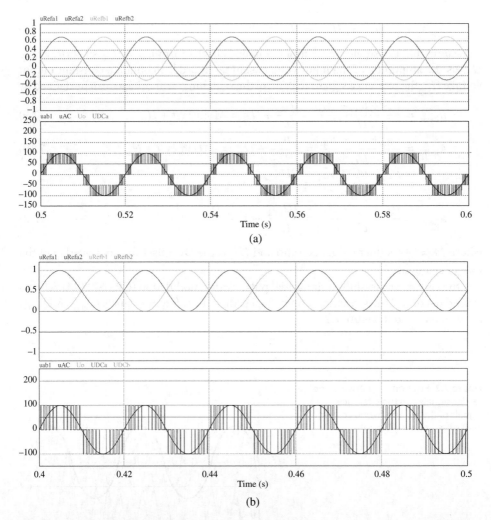

Figure 7.3 Simulation waveforms of the proposed six-arm hybrid converter with single-phase AC input. (a) Equal DC inputs. (b) Different DC inputs.

It is obvious that U_{os} should satisfy

$$\begin{cases} U_{os} + M < 1 \\ U_{os} - M > \max(u_{Refa2}, u_{Refb2}) \end{cases} \tag{7.13}$$

and we have

$$\max(u_{Refa2}, \ u_{Refb2}) + M < 1 - M \tag{7.14}$$

The output voltage U_O should be selected in the range

$$U_O > \max(U_{DCa}, \ U_{DCb}) + U_{AC} \tag{7.15}$$

7.2.3 Simulation Verification

In this section, the feasibility of the proposed six-arm hybrid converter will be proven by the simulation results. As the dual DC input voltages can be equal or different, two sets of simulation parameters are listed in Table 7.1, in which U_{os} is selected based on Eq. (7.13).

By setting $N = 4$, the simulation waveforms of both cases are shown in Figure 7.3. It is found that the reference signals defined by Eqs. (7.9)–(7.12) do not overlap; the difference between u_{a1} and u_{b1} (that is u_{ab1}) has the same shape as the AC input voltage u_{AC}, and the output voltage U_O remains at the desired value 200 V ($U_O = 200$ V). Therefore, the proposed converter can convert dual DC input sources with identical or different voltages and one single-phase AC source to constant DC output voltage.

7.3 Nine-Arm Hybrid Converter with Three-Phase AC Input

7.3.1 Basic Structure and Operating Principle

If the AC input voltage source of the hybrid converter is three-phase, based on the nine-arm AC–DC converter in Section 4.4, the nine-arm hybrid converter with three-phase AC input can be obtained as in Figure 7.4, by connecting one group of endpoints (e.g. u_2, v_2, and w_2) to three DC voltage sources [2].

Similar to Section 7.2.1, the relationships between the input voltages and the switching-arm voltages are listed as follows:

$$\begin{cases} u_{iu} - u_{iv} = u_{u1} - u_{v1} = \dfrac{1}{2}[(u_{U3} + u_{U2} - u_{U1}) - (u_{V3} + u_{V2} - u_{V1})] \\ u_{iv} - u_{iw} = u_{v1} - u_{w1} = \dfrac{1}{2}[(u_{V3} + u_{V2} - u_{V1}) - (u_{W3} + u_{W2} - u_{W1})] \\ u_{iw} - u_{iu} = u_{w1} - u_{u1} = \dfrac{1}{2}[(u_{W3} + u_{W2} - u_{W1}) - (u_{U3} + u_{U2} - u_{U1})] \end{cases} \tag{7.16}$$

$$\begin{cases} U_{DCu} = u_{u2} = \dfrac{U_O + u_{U3} - u_{U2} - u_{U1}}{2} \\ U_{DCv} = u_{v2} = \dfrac{U_O + u_{V3} - u_{V2} - u_{V1}}{2} \\ U_{DCw} = u_{w2} = \dfrac{U_O + u_{W3} - u_{W2} - u_{W1}}{2} \end{cases} \tag{7.17}$$

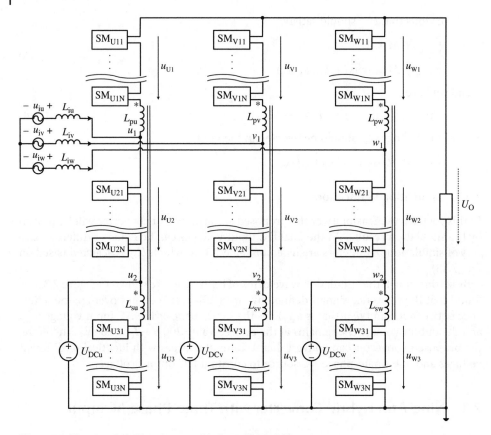

Figure 7.4 Nine-arm hybrid converter with three-phase AC input.

where u_{X1}, u_{X2}, and u_{X3} are the switching arm voltages of phase unit X (X = U, V, W). The desired DC output voltage U_O can be obtained by controlling the switching-arm voltages.

7.3.2 Control Scheme

Similar to Sections 4.4.2 and 7.2.2, the reference signal related to the input AC voltage source is defined by

$$u_{\text{Refx1}} = \frac{u_{X3} + u_{X2} - u_{X1}}{U_O} = \frac{2u_{ix}}{U_O} + U_{os} = M_i \sin(2\pi ft + \varphi + \phi_x) + U_{os} \qquad (7.18)$$

where $x = u, v, w$, $X = U, V, W$, $\phi_u = 0$, $\phi_v = -\frac{2\pi}{3}$, $\phi_w = \frac{2\pi}{3}$. $M_i = \frac{2U_i}{U_O}$ is the modulated ratio, while U_i, f, and φ are the phase voltage amplitude, frequency, and phase shift of the input voltage, respectively. U_{os} is the DC offset of the reference signal.

Table 7.2 Parameters of the nine-arm hybrid converter with three-phase AC input.

Case	U_i (V)	f (Hz)	ϕ (rad)	U_{DCu} (V)	U_{DCv} (V)	U_{DCw} (V)	U_{os} (V)	U_O (V)	Figure
1	70	50	0	50	50	50	0.25	200	7.5
2	70	50	0	60	50	40	0.3	200	7.6

The reference signals related to the triple-input DC voltage sources are in the form

$$u_{Refx2} = \frac{u_{X3} - u_{X2} - u_{X1}}{U_O} = \frac{2U_{DCx}}{U_O} - 1 \tag{7.19}$$

In order to avoid overlap of the reference signals, U_{os} should satisfy

$$\max(u_{Refx2}) + M_i < U_{os} < 1 - M_i \tag{7.20}$$

and the desired output voltage U_O should be selected in the range

$$U_O > \max(U_{DCu}, \ U_{DCv}, \ U_{DCw}) + 2U_i \tag{7.21}$$

7.3.3 Simulation Verification

Two sets of simulation parameters listed in Table 7.2 are used to prove the feasibility of the proposed nine-arm hybrid converter, in which the DC voltage sources have equal values in Case 1 and different ones in Case 2. The DC offset of the reference signal U_{os} and the output voltage U_O are selected based on Eqs. (7.20) and (7.21), respectively. By setting $N = 4$, the simulation waveforms of input voltages, output voltage, switching-arm voltages, and reference signals are shown in Figures 7.5 and 7.6.

It is found that the output voltage U_O remains at the desired value 200 V in both cases ($U_O = 200$ V). The combination of switching-arm voltage $u_{U3} + u_{U2} - u_{U1}$ has the same phase and double fundamental component as the input phase voltage u_{iu}, while the average value of $u_{U3} - u_{U2} - u_{U1}$ is equal to $2U_{DCu} - U_O$, which agrees with the theoretical analysis. Therefore, the proposed converter can output a constant DC voltage with both DC and AC voltage sources.

7.4 Multiple-Arm Hybrid Converter

7.4.1 Basic Structure and Operating Principle

If the number of input voltage sources increases, then the multiple-arm hybrid converters could be constructed by adding some switching arms in the hybrid converters in Figures 7.1 and 7.4, according to the multiple-input AC–DC converter proposed in Chapter 4 and the multiple-terminal DC–DC converter proposed in Chapter 6. As there

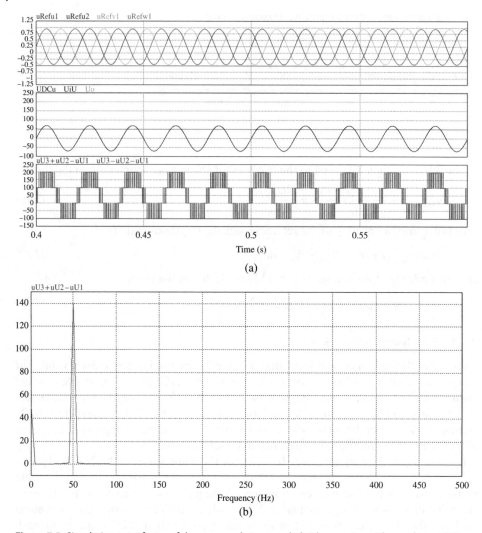

Figure 7.5 Simulation waveforms of the proposed nine-arm hybrid converter with equal input DC sources. (a) Waveforms. (b) Spectrum of $u_{U3} + u_{U2} - u_{U1}$.

are different ways to connect the AC and DC voltage sources to the switching arm terminals, only the multiple-arm hybrid converter with $M - 2$ AC inputs will be discussed in this section, where M is the number of switching arms in one general phase unit. The other types of hybrid converter can be analyzed in a similar way.

The proposed multiple-arm hybrid converters with single-phase AC input and three-phase AC input are illustrated in Figures 7.7 and 7.8, respectively. According to Chapter 4, the terminal voltage of the switching arm is determined by

$$u_{xk} = \frac{1}{2}\left(U_O - \sum_{i=1}^{k} u_{Xi} + \sum_{j=k+1}^{M} u_{Xj} \right) \tag{7.22}$$

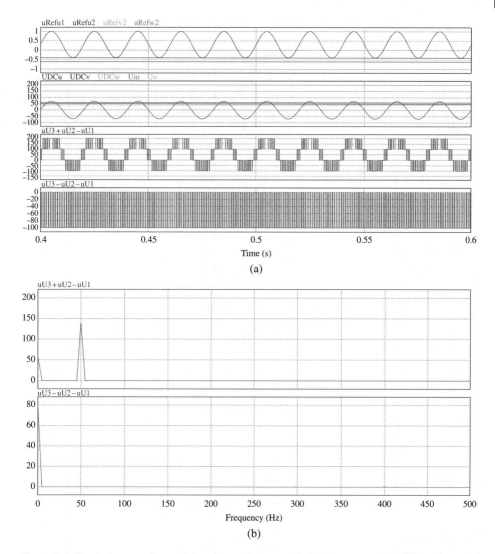

Figure 7.6 Simulation waveforms of the proposed nine-arm hybrid converter with different input DC sources. (a) Waveforms. (b) Spectrums of $u_{U3} + u_{U2} - u_{U1}$ and $u_{U3} - u_{U2} - u_{U1}$.

where $k = 1, 2, \ldots, M - 1$, $x = a, b$, and $X = A, B$ for the case with single-phase AC input; $x = u, v, w$ and $X = U, V, W$ for that with three-phase AC input.

As shown in Figures 7.7 and 7.8, the voltage between the switching arms connected to the AC voltage sources can be expressed by

$$
\begin{aligned}
u_{xk} - u_{yk} &= \frac{1}{2} \left(-\sum_{i=1}^{k} u_{Xi} + \sum_{j=k+1}^{M} u_{Xj} + \sum_{i=1}^{k} u_{Yi} - \sum_{j=k+1}^{M} u_{Yj} \right) \\
&= \begin{cases} u_{ACk}, & \text{for single-phase} \\ u_{ACkx} - u_{ACky}, & \text{for three-phase} \end{cases}
\end{aligned}
\tag{7.23}
$$

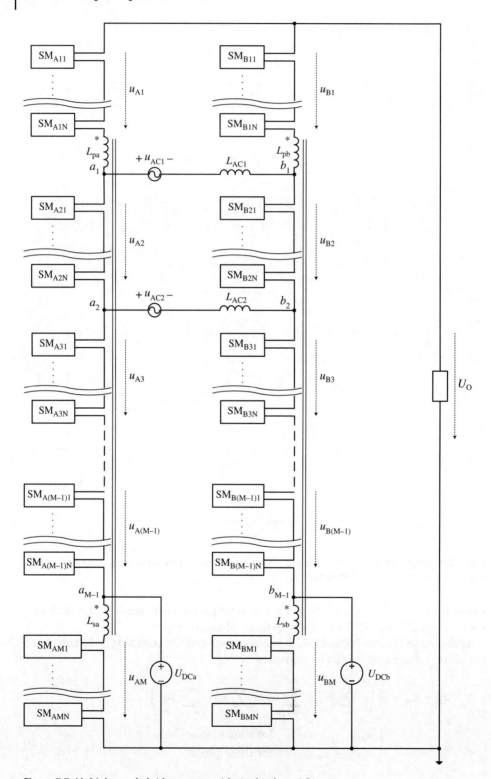

Figure 7.7 Multiple-arm hybrid converter with single-phase AC input.

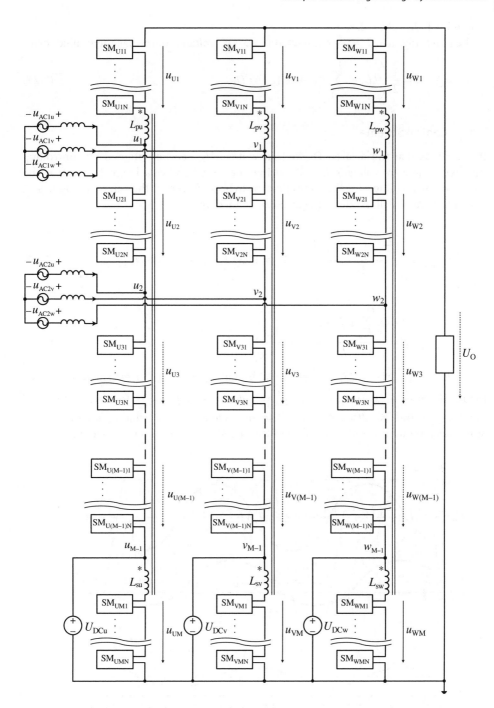

Figure 7.8 Multiple-arm hybrid converter with three-phase AC input.

where $k = 1, 2, \ldots, M-2$.

The switching-arm voltage connected to the DC voltage source has the unified form

$$u_{x(M-1)} = \frac{1}{2}\left(U_O - \sum_{i=1}^{M-1} u_{Xi} + u_{XM}\right) = U_{DCx} \quad (7.24)$$

7.4.2 Control Scheme

It is obvious that the desired DC output voltage U_O can be obtained by controlling the switching-arm voltages u_{Xi}. By using the CPS-SPWM scheme to control the switching-arm voltages of the multiple-arm hybrid converter, the reference signals of one phase unit are illustrated in Figure 7.9, defined by

$$u_{Refkx} = \frac{-\sum\limits_{i=1}^{k} u_{Xi} + \sum\limits_{j=k+1}^{M} u_{Xj}}{U_O} = M_k \sin(2\pi f_k t + \varphi_k + \phi_{kx}) + U_{osk} \quad (7.25)$$

and

$$u_{Ref(M-1)k} = \frac{-\sum\limits_{i=1}^{M-1} u_{Xi} + u_{XM}}{U_O} = \frac{2U_{DCx}}{U_O} - 1 \quad (7.26)$$

where $k = 1, 2, \ldots, M-2$, $M_{kx} = \frac{U_{ACk}}{U_O}$ ($x = a, b$) and $\phi_{ka} = 0$, $\phi_{kb} = \pi$ for single-phase AC input, $M_{kx} = \frac{2U_{ACk}}{U_O}$ ($x = u, v, w$) and $\phi_{ku} = 0$, $\phi_{kv} = -\frac{2\pi}{3}$, $\phi_{kw} = \frac{2\pi}{3}$ for three-phase AC input. U_{osk} is the DC offset of the kth reference signal. U_{ACk}, f_k, and φ_k are the voltage amplitude, frequency, and phase shift of the kth AC voltage source, respectively.

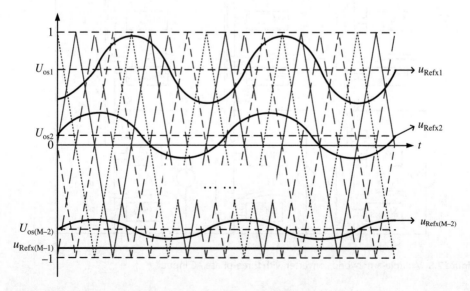

Figure 7.9 Relationship between the carrier signals and the reference signals of the multiple-arm hybrid converter.

As the DC offsets of the reference signals should satisfy the following requirements:

$$
\begin{cases}
U_{os1} + M_1 < 1 \\
U_{os2} + M_2 < U_{os1} - M_1 \\
\ldots \\
U_{osk} + M_k < U_{os(k-1)} - M_{(k-1)} \\
\ldots \\
\max(u_{Ref(M-1)k}) < U_{os(k-2)} - M_{(k-2)}
\end{cases}
\tag{7.27}
$$

the desired output voltage U_O should be selected in the range

$$
\begin{cases}
U_O > \max(U_{DCa}, U_{DCb}) + \displaystyle\sum_{k=1}^{M-2} U_{ACk}, & \text{for single-phase} \\
U_O > \max(U_{DCu}, U_{DCv}, U_{DCw}) + 2\displaystyle\sum_{k=1}^{M-2} U_{ACk}, & \text{for three-phase}
\end{cases}
\tag{7.28}
$$

When M = 3, Eq. (7.28) is consistent with Eq. (7.15) or Eq. (7.21), which verifies the feasibility of the proposed multiple-arm hybrid converter.

7.5 Summary

Compared with the traditional approach that uses several independent converters to connect different kinds of input to the high-voltage DC bus, a hybrid high-voltage converter is provided for the first time, which can convert multiple AC and DC inputs to single DC output with only one converter. Similar to the multi-terminal high-voltage converters presented in Chapters 3–6, the proposed converter still keeps the advantage of simple structure and fewer components, and is particularly suitable for the occasion with various voltage sources existing.

References

1 Zhang, B., Fu, J., Qiu, D. Y. Six switching-groups MMC hybrid converter and its control method. State Intellectual Property Office of the P.R.C., ZL 201410134645.4, 2016.10.5.

2 Zhang, B., Fu, J., Qiu, D. Y. Nine switching-groups MMC hybrid converter and its control method. State Intellectual Property Office of the P.R.C., ZL 201410134601.1, 2016.10.5.

8

Short-Circuit Protection for High-voltage Converters

8.1 Introduction

The modular multilevel converter (MMC) has become the most attractive converter topology for high-voltage direct-current (HVDC) transmission systems because of its modularity and scalability. Normally, the DC side of the MMC is connected to the DC transmission line, while its AC side is connected to a utility grid through a three-phase transformer. Thus, one of the major challenges associated with MMC-HVDC systems is the capability to handle DC-side short-circuit faults.

For the MMC with conventional half-bridge sub-modules (HBSMs) shown in Figure 1.10a, the driving signals of the switches will be blocked when a DC-side short-circuit fault happens, then all of the HBSMs operate in the "Energization" state listed in Table 1.5. The fault current, as shown in Figure 8.1, flows from the AC side toward the DC side through the antiparallel diodes of the HBSMs, and its amplitude is determined by the AC-side system parameters, arm inductors, DC transmission lines/cables characteristics, and the fault location [1]. In this case, the DC short-circuit current is very large and cannot extinguish the arc, which will easily damage the device and also bring system outage. This problem is particularly prominent in overhead transmission lines.

Expect for using conventional AC circuit breakers (CBs) and fuses to protect HVDC systems on the AC side, the existing solutions to interrupt and clear the DC-side short-circuit fault of an MMC-HVDC system with HBSMs can be summarized as follows [1]: (i) employ a DC-side CB; (ii) use sub-modules (SMs) with DC fault-handling capability in the switching arms, such as the full-bridge sub-module (FBSM); (iii) insert FBSMs into the phase leg, DC side or AC side, of the MMC-HVDC system.

Similar to MMC, the multi-terminal high-voltage AC–DC or DC–AC converters presented in Chapters 3 and 4 are also composed of SMs. The DC blocking capability should be considered when the proposed multi-terminal high-voltage converters are applied on HVDC occasions. According to the DC fault solutions of the MMC-HVDC system, three short-circuit protection schemes will be discussed for the proposed multi-terminal high-voltage converters in the rest of this chapter. First, a novel modular DC CB which makes use of the structure of the switching arm is introduced in Section 8.2. Then, several typical kinds of SM with DC fault-handling capability are analyzed in Section 8.3, which can replace HBSMs in the multi-terminal high-voltage converters. Finally, the feasible architectures of a hybrid multi-terminal high-voltage converter with short-circuit protection feature are proposed in Section 8.4.

Multi-terminal High-voltage Converter, First Edition. Bo Zhang and Dongyuan Qiu.
© 2019 John Wiley & Sons Singapore Pte. Ltd. Published 2019 by John Wiley & Sons Singapore Pte. Ltd.

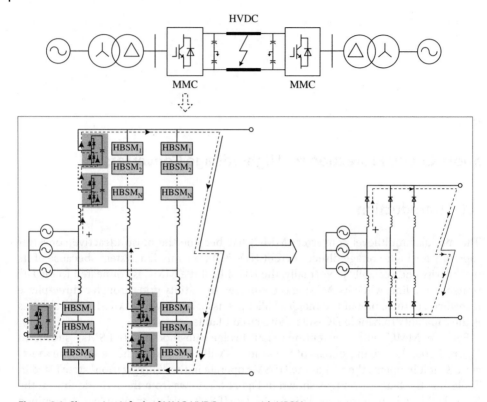

Figure 8.1 Short-circuit fault of MMC-HVDC system with HBSMs.

8.2 Modular DC Circuit Breaker

With the rapid development of HVDC transmission, there is a growing demand for DC CBs. The difficulties of realizing HVDC CBs can be attributed to the demanding require-ments on CBs in DC systems, which are quite different from those of AC CBs. One of the major differences is the absence of natural zero-current crossings in DC systems. The other is the small impedance of DC systems, which will cause the fault current to increase rapidly.

At present, the main types of high-voltage DC CBs include mechanical CBs, solid-state CBs, and hybrid CBs. Mechanical DC CBs can be regarded as adding an oscillatory commutation circuit based on the conventional mechanical AC CB, which can automatically form the zero crossing of high-frequency oscillating current during breaking of the DC current. However, the blocking capacity of a mechanical DC CB is limited; the cutoff time of fault currents is relatively long, and the mechanical contacts are easily damaged by on–off arcing. Solid-state DC CBs utilize the blocking capability of the semi-controlled or full-controlled power electronic switching devices, which have the advantages of accurately controllable switching time, long life, and so on. However, power electronic switches easily tend to be over-voltage or over-current, and high conduction losses of the switching devices result in a bulky cooling system. A hybrid DC CB is a combination of a mechanical CB and a solid-state CB, and has all of their advantages, such as low on-state loss, rapid and controllable breaking, no arcing, and no need for special cooling equipment. The main challenge of hybrid DC CBs is how to divert the current precisely from the mechanical switch to the power electronic switch.

In this section, a novel hybrid DC CB named the modular DC CB is proposed [2], which consists of a switching arm, diode D, resistor R, and mechanical switch S. As shown in Figure 8.2, the aforementioned switching arm is composed of N HBSMs, where N is a positive integer. Assume that the terminal voltage is U_b, then the maximum voltage of every insulated-gate bipolar transistor (IGBT) or capacitor in the HBSM is only U_b/N. The larger N is, the lower the voltage stress on the switching device.

Taking N = 2 for example, the equivalent circuit of the proposed modular DC CB is shown in Figure 8.3a. During normal operating state, the mechanical switch S stays in the closed position and the current flows through S, as shown in Figure 8.3b.

When a DC short-circuit fault occurs, the fault current instantly increases and flows through the mechanical switch S. The DC fault-clearing and reset process of the proposed modular DC CB is divided into the following six modes.

Mode 1 (Figure 8.4a): If the current flowing through S is greater than a set value, then the mechanical switch S opens and an arc is established between the metallic contacts of

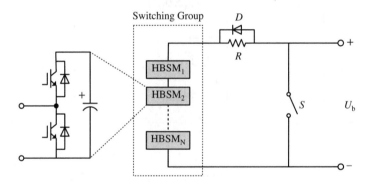

Figure 8.2 Topology of modular DC circuit breaker.

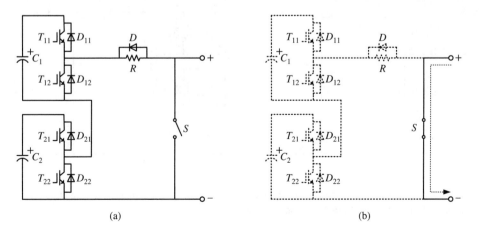

Figure 8.3 Modular DC circuit breaker with N = 2. (a) Equivalent circuit. (b) Normal operating mode.

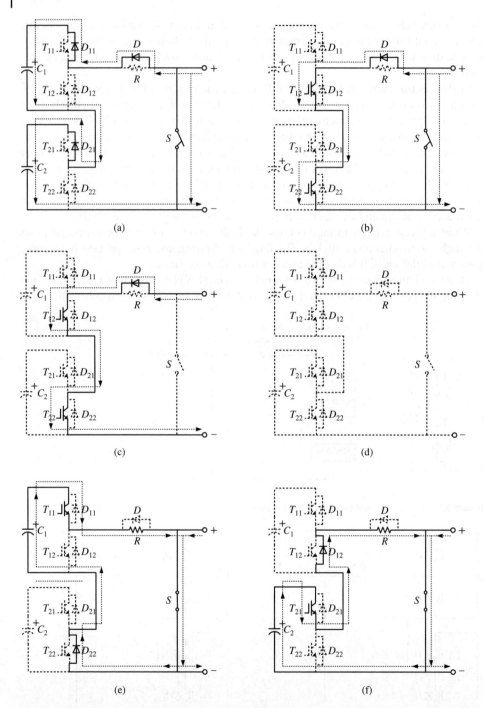

Figure 8.4 Operating process of modular DC CB when a DC short-circuit fault happens. (a) Mode 1. (b) Mode 2. (c) Mode 3. (d) Mode 4. (e) Mode 5. (f) Mode 6.

the switch. At the same time, the fault current flows through diode D and the antiparallel diodes and capacitors of the HBSMs, that is, D_{11}, C_1, D_{21}, and C_2.

Mode 2 (Figure 8.4b): As the turning-on signal has been sent to the second IGBT of the HBSM in Mode 1, T_{12} and T_{22} will conduct when both C_1 and C_2 are charged to a sufficiently high voltage. Then, the fault current flows through T_{12} and T_{22} instead of D_{11} and D_{21}, resulting in a reduced current flowing through the mechanical switch S.

Mode 3 (Figure 8.4c): The arc between the mechanical switch contacts is extinguished when the mechanical switch current is decreased to a low value. At this moment, the mechanical switch S is completely disconnected.

Mode 4 (Figure 8.4d): By turning T_{12} and T_{22} off, the DC short-circuit fault is consequently blocked.

Mode 5 (Figure 8.4e): The mechanical switch S is turned on again after the DC short-circuit fault is cleared. The turning-on signal is sent to T_{11}, the first IGBT of HBSM$_1$, then C_1 is discharged through the loop T_{11}–R–S–D_{22}. Mode 5 ends when the capacitor voltage of C_1 is equal to zero, and T_{11} is turned off.

Mode 6 (Figure 8.4f): Similar to Mode 5, T_{21} is turned on to discharge C_2. In this way, the capacitor discharge can be controlled accurately, and the discharging current can be reduced effectively because the capacitors are discharged one after another.

The proposed modular DC CB returns to its normal operating mode as shown in Figure 8.3b after all the HBSMs have been discharged. Obviously, the modular structure, low voltage stress, small discharging current, and accurate control are advantages of the proposed modular DC CB.

8.3 Sub-Modules with DC Fault-Handling Capability

8.3.1 Full-Bridge Sub-Module

The FBSM, as shown in Figure 8.5a, is composed of four IGBTs with their antiparallel diodes and one capacitor. When all of the IGBTs of the HBSMs are turned OFF, the HBSMs operate in the "Energization" states and the capacitors of the HBSMs are inserted into the current loops, as shown in Figure 8.5b,c.

If all SMs used in the MMC-HVDC system are realized by FBSMs, subsequent to a DC-side short current fault, each leg turns out to be diodes in series with capacitors, as shown in Figure 8.6. As the capacitor voltages can generate reverse voltages to block the short-circuit currents, FBSMs can provide DC fault-handling capability. Though the

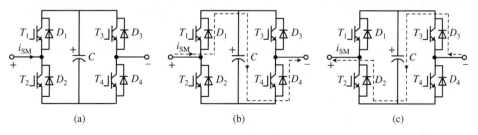

(a)   (b)   (c)

Figure 8.5 Full-bridge sub-module and its current path when all switches are OFF. (a) Schematic. (b) $i_{SM} > 0$. (c) $i_{SM} < 0$.

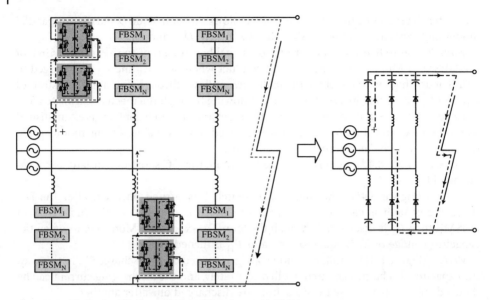

Figure 8.6 Short-circuit fault of MMC-HVDC system with FBSMs.

power losses as well as the cost of the system using FBSMs are significantly higher than those using HBSMs, the multi-terminal high-voltage converter with FBSMs eliminates the current path of the freewheeling diodes and can potentially interrupt the fault current within a very short time.

8.3.2 Clamp-Double Sub-Module

Because a reverse voltage polarity at the DC bus is not required for HVDC applications, only positive voltage should be inserted in normal operation. Therefore, a clamp-double sub-module (CDSM), shown in Figure 8.7, is considered as an advantageous realization in MMC-HVDC systems [3], enabling the desired cutoff and voltage clamping functionality.

A CDSM consists of two HBSMs, two additional diodes, and one extra IGBT with its antiparallel diode. During normal operation, the switch T_5 is normally ON and the CDSM represents two series-connected HBSMs. As shown in Figure 8.8, the capacitors can be bypassed by turning T_2 and T_3 ON and inserted by turning T_1 and T_4 ON. Assuming that both capacitor voltages are equal to U_C, the output voltage of the CDSM will be $2U_C$. It is found that the total expense for the semiconductors and the resulting

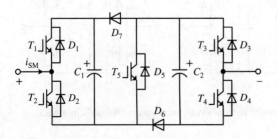

Figure 8.7 Topology of clamp-double sub-module.

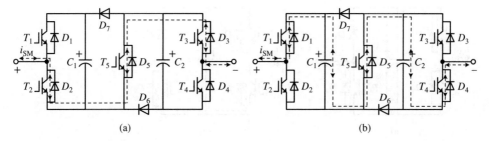

Figure 8.8 Current paths of clamp-double sub-module in normal operation. (a) Capacitors bypassed. (b) Capacitors inserted.

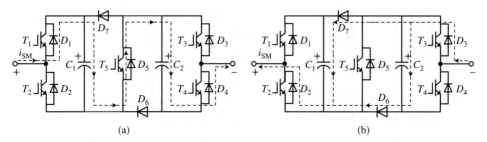

Figure 8.9 Current paths of clamp-double sub-module when all switches are OFF. (a) $i_{SM} > 0$. (b) $i_{SM} < 0$.

losses is only slightly increased because of the extra switch T_5. However, compared to the half and full-bridge MMCs with the same number of voltage levels, the loss of a clamp-double MMC is higher than that of a half-bridge MMC and lower than that of a full-bridge MMC.

In order to cut off the fault currents, T_5 as well as four other switches must be blocked. As shown in Figure 8.9a, capacitors C_1 and C_2 are connected in series when the arm current is positive or $i_{SM} > 0$. When $i_{SM} < 0$, C_1 and C_2 are connected in parallel, as shown in Figure 8.9b, and the reversed voltage is half that produced by the FBSM.

8.3.3 Unipolar-Voltage Sub-Module

The FBSM is well known for its simple method of cutting off the fault currents in any direction, but some additional switching states are not very useful in normal operation. Thus, the topology of a unipolar-voltage sub-module (UVSM), shown in Figure 8.10, is modified from the FBSM, in which one IGBT and its antiparallel diode are replaced by a diode [1]. The equivalent circuits during normal operation of the UVSM are shown

Figure 8.10 Topology of unipolar-voltage sub-module.

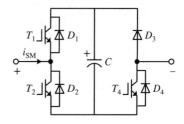

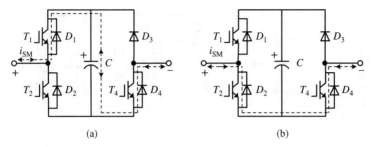

Figure 8.11 Current paths of unipolar-voltage sub-module in normal operation. (a) Capacitor inserted. (b) Capacitor bypassed.

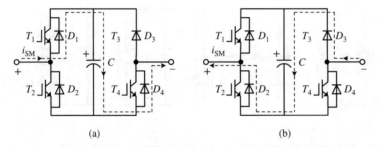

Figure 8.12 Current paths of unipolar-voltage sub-module when all switches are OFF. (a) $i_{SM} > 0$. (b) $i_{SM} < 0$.

in Figure 8.11; while switch T_4 is always on, the capacitor C is inserted or bypassed by turning on switch T_1 or T_2. During a DC-side short-circuit fault, the current paths of the UVSM are shown in Figure 8.12, which are similar to those of the FBSM. Therefore, the UVSM has the ability to block DC fault currents.

8.3.4 Cross-Connected Sub-Module

In order to reduce the cost and loss of the sub-module, a cross-connected sub-module (CCSM), which can produce more voltage levels with the least number of device requirements, is proposed in Figure 8.13. Here, two HBSMs are connected back-to-back in a crossed way using two extra IGBTs with their antiparallel diodes [4].

Since two HBSMs can be controlled independently, only one capacitor or both capacitors can be inserted. Assuming that both capacitor voltages are the same as U_C,

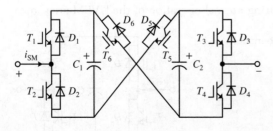

Figure 8.13 Topology of cross-connected sub-module.

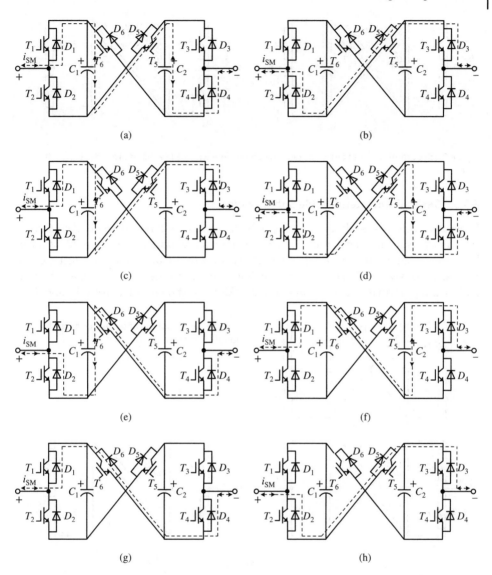

Figure 8.14 Current paths of cross-connected sub-module in normal operation. (a) Insert two capacitors and output $2U_C$. (b) Insert two capacitors and output $-2U_C$. (c) Insert C_1 and output U_C. (d) Insert C_2 and output U_C. (e) Insert C_1 and output $-U_C$. (f) Insert C_2 and output $-U_C$. (g) Capacitors bypassed by T_1, T_4, T_6, and output 0 V. (h) Capacitors bypassed by T_2, T_3, T_5, and output 0 V.

the five-level output voltages of $\{-2U_C, -U_C, 0, U_C, 2U_C\}$ can be generated in normal operation according to the different switching states of the CCSM. All corresponding current paths are shown in Figure 8.14; it is found that any voltage level can be achieved in either current direction, which can offer more flexibility in capacitor voltage control.

In short-circuit circumstances, two capacitors of the CCSM are connected in series to block the short-circuit current, as shown in Figure 8.15.

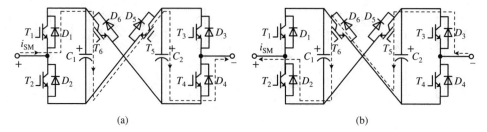

Figure 8.15 Current paths of cross-connected sub-module when all switches are OFF. (a) $i_{SM} > 0$. (b) $i_{SM} < 0$.

8.3.5 Series-Connected Double Sub-Module

If negative voltages, such as $-U_C$ and $-2U_C$, are not required in some applications, then a novel sub-module named the series-connected double sub-module (SDSM) can be used. As shown in Figure 8.16, the SDSM is derived from the CCSM using diode D_6 instead of switch T_6 [1], and the possible switching states and equivalent current paths during normal operation and fault blocking are shown in Figures 8.17 and 8.18,

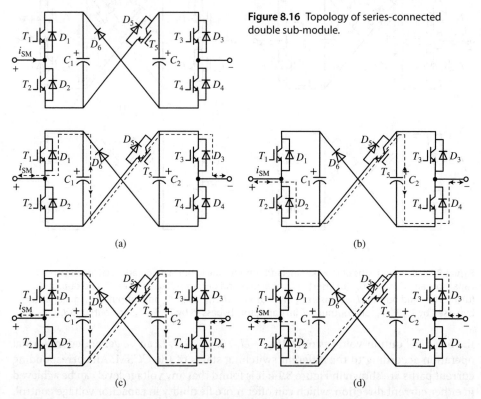

Figure 8.16 Topology of series-connected double sub-module.

Figure 8.17 Current paths of series-connected double sub-module in normal operation. (a) Capacitor C_1 inserted. (b) Capacitor C_2 inserted. (c) Capacitors C_1 and C_2 inserted. (d) Capacitors bypassed.

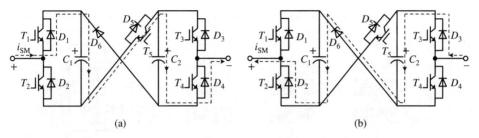

(a)   (b)

Figure 8.18 Current paths of series-connected double sub-module when all switches are OFF. (a) $i_{SM} > 0$. (b) $i_{SM} < 0$.

respectively. During normal operation, the conduction switch T_5 is always on, while D_6 is biased to off, so the SDSM can be considered as two independent HBSMs connected in series. Compared with the CDSM, only one diode (D_6) is needed in the SDSM, but one more voltage level (U_C) can be produced at the output. Furthermore, similar to the CCSM, both capacitors in the SDSM remain in series and charged regardless of the fault current direction.

8.4 Configuration of the Hybrid Multi-terminal High-voltage Converter

By using the SMs with DC fault-handling capability presented in Section 8.3, the proposed multi-terminal high-voltage converter can suppress the DC link fault current without tripping the AC CB. Although the HBSM cannot interrupt the fault current, it provides the highest efficiency and the lowest cost, due to its lowest number of components. Thus, schemes based on a single type of SM circuit configuration, such as FBSMs, CDSMs, or CCSMs, do manage the fault condition, but with the sacrifice of cost and efficiency. To pursue the optimal design in terms of DC fault-handling capability, efficiency, and cost, a hybrid design concept making use of different features of various SMs has been put forward [1]. In this section, three kinds of hybrid multi-terminal high-voltage converter composed of HBSMs and FBSMs are provided, that is, HBSMs for lower losses in conjunction with FBSMs for DC fault-handling capability.

As shown in Figure 8.19a, every switching arm of a hybrid multi-terminal AC–DC converter consists of N HBSMs and M FBSMs, where N and M represent the number of series-connected HBSMs and FBSMs, respectively. Subsequent to fault detection, all IGBTs are turned off; M capacitors per switching arm will be inserted in the short-circuit loop, producing an opposing voltage to block the fault current. Thus, the fault-blocking time is determined by M; the higher M is, the higher the reverse voltage generated.

Since the DC fault current flows from the AC side toward the DC side of the multi-terminal AC–DC converter, MMC-HVDC systems with DC fault-handling capacity can be constructed by connecting some FBSMs in series on the AC side or the DC side [5, 6]. Similarly, in addition to embedding FBSMs into the switching arms, FBSMs can be inserted in the AC-side or DC-side buses of the multi-terminal AC–DC converter, as illustrated in Figure 8.19b,c, respectively.

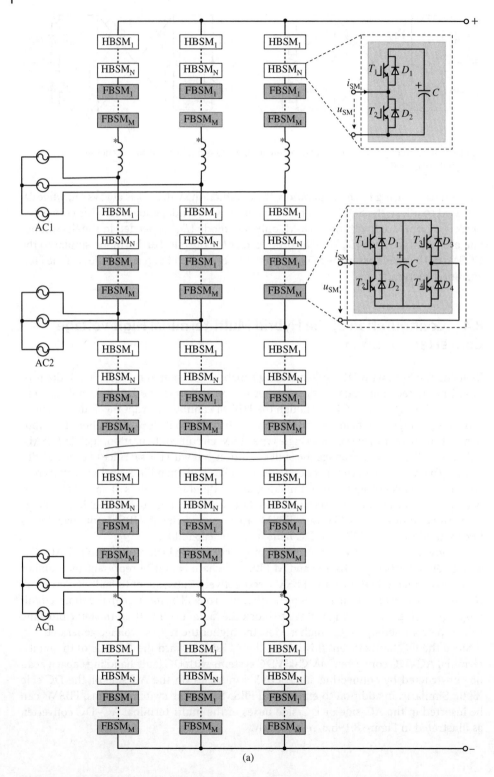

(a)

Figure 8.19 Block diagrams of hybrid multi-terminal high-voltage converter. (a) FBSMs on the switching arm. (b) FBSMs on the AC side. (c) FBSMs on the DC side.

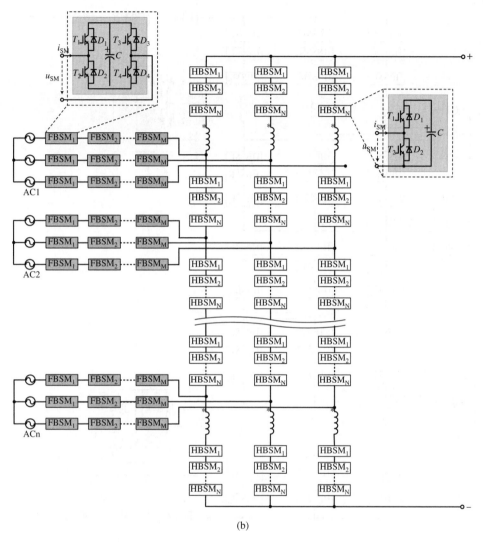

(b)

Figure 8.19 (*Continued*)

In this way, the hybrid multi-terminal high-voltage converter can potentially offer lower losses and costs, as well as the DC fault-handling capability. Evaluation and comparison among different configurations of hybrid multi-terminal high-voltage converter should be carried out in terms of DC fault-handling capability, semiconductor power losses, and component requirements.

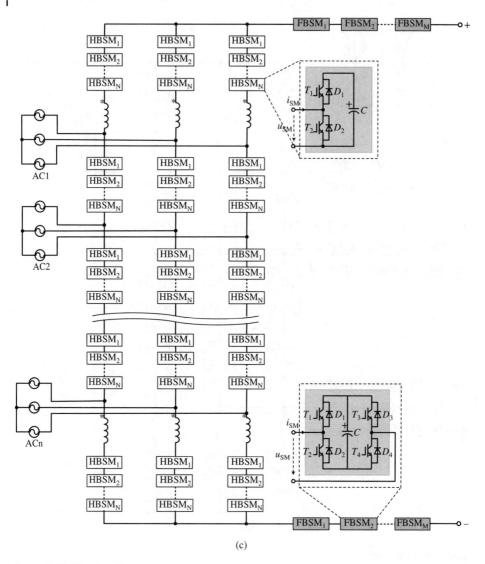

(c)

Figure 8.19 (*Continued*)

8.5 Summary

Similar to any other voltage-source converter applied in an HVDC system, the proposed multi-terminal AC–DC and DC–AC converters with HBSMs lack the capability to handle DC-side short-circuit faults. In this chapter, three short-circuit protection schemes have been investigated for multi-terminal high-voltage converters, including developing a modular DC CB, using a single type of SM with DC fault-handling capability, and adopting a hybrid configuration based on a combination of HBSM and FBSM. Therefore, the multi-terminal high-voltage converter with DC fault-current limitation can meet the demanding future applications in power transmission.

References

1 Qin, J., Saeedifard, M., Rockhill, A. et al. (2015). Hybrid design of modular multilevel converters for HVDC systems based on various submodule circuits. *IEEE Transactions on Power Delivery* 30 (1): 385–394.
2 Zhang, B., Fu, J., Qiu, D. Y. Modular high-voltage DC circuit breaker. State Intellectual Property Office of the P.R.C., ZL 201410191580.7, 2017.8.25.
3 Marquardt, R. (2010) Modular multilevel converter: a universal concept for HVDC-networks and extended DC-bus-applications. The 2010 International Power Electronics Conference (ECCE ASIA), pp. 502–507.
4 Nami, A., Wang, L., Dijkhuizen, F., A. Shukla (2013) Five level cross connected cell for cascaded converters. 15th European Conference on Power Electronics and Applications (EPE), Lille, France, pp. 1–9.
5 Zhang, B., Fu, J., Qiu, D. Y. Hybrid modular multilevel converter with cascaded H-bridge on ac side. State Intellectual Property Office of the P.R.C., ZL 201420520728.2, 2015.3.28.
6 Zhang, B., Fu, J., Qiu, D. Y., Han, C. Hybrid modular multilevel converter with cascaded H-bridge on dc side. State Intellectual Property Office of the P.R.C., ZL 201520741909.2, 2016.3.2.

9

Common Techniques and Applications of Multi-terminal High-voltage Converters

9.1 Introduction

Similar to the structure of the modular multilevel converter (MMC), the multi-terminal high-voltage converters proposed in Chapters 3–7 are stacked up with a large number of identical sub-modules (SMs). Thus, the voltage control strategy to balance and maintain the capacitor voltages of SMs is also needed for multi-terminal high-voltage converters. The voltage balancing strategies for MMCs can generally be summarized as two types: one the sorting method [1] and the other the closed-loop control [2]. In the sorting method, the SM capacitor voltages of each arm are monitored and sorted. When the arm current is positive, a certain number of SMs with the lowest voltages are identified and switched on or inserted. When the arm current is negative, the SMs with the highest voltages are identified and switched on. In the closed-loop control, the capacitor voltage of each SM is sampled and compared with the command capacitor voltage, then the charging and discharging time of each capacitor is decided to make the individual capacitor voltage follow its command.

Based on the closed-loop control method, the capacitor voltage balancing schemes for several multi-terminal high-voltage converters are presented in detail in this chapter. Moreover, the possible applications of multi-terminal high-voltage converters in renewable energy sources and variable-frequency systems are proposed.

9.2 Capacitor Voltage Control Scheme for Multi-terminal High-voltage Converters

9.2.1 Single-Input Dual-Output DC–AC Converter

In this section, the half-bridge single-phase DC–AC converter shown in Figure 3.1, the full-bridge single-phase DC–AC converter shown in Figure 3.6, and the three-phase DC–AC converter shown in Figure 3.9 are selected to introduce the capacitor voltage control scheme for single-input dual-output converters. Since the control schemes presented in Sections 3.2.2, 3.3.3, and 3.4.2 are only available to generate two AC outputs, the reference voltage for each arm should comprise an additional closed-loop controlling signal to balance all capacitor voltages.

The schematic to generate the voltage balancing control signal $\Delta u^*_{\mathrm{XCi}}$ (where X = U, M, L) for the half-bridge single-phase DC–AC converter is shown in Figure 9.1 [3].

Multi-terminal High-voltage Converter, First Edition. Bo Zhang and Dongyuan Qiu.
© 2019 John Wiley & Sons Singapore Pte. Ltd. Published 2019 by John Wiley & Sons Singapore Pte. Ltd.

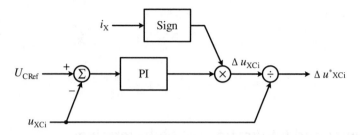

Figure 9.1 Voltage balancing control signal for the single-input dual-output half-bridge single-phase DC–AC converter.

First, the SM capacitor voltage u_{XCi} is sampled and subtracted from the reference capacitor voltage U_{CRef}, and their difference value is sent to a proportional integral (PI) controller. Then, the voltage error signal Δu_{XCi} is generated by multiplying the output of the PI controller by the sign function of the corresponding arm current i_X, which is equal to 1 when i_X is positive or -1 when i_X is negative. Finally, the voltage balancing signal Δu^*_{XCi} is obtained by normalizing Δu_{XCi}.

Based on Eqs. (3.6) and (3.9) and the control strategy in Figure 3.4, the reference signals for each arm when considering the capacitor voltage control are defined by the following equations:

$$\begin{cases} u_{Ui,Ref} = u_{Ref1} + \Delta u^*_{UCi} \\ u_{Mi,Ref1} = u_{Ref1} + \Delta u^*_{MCi} \\ u_{Mi,Ref2} = u_{Ref2} + \Delta u^*_{MCi} \\ u_{Li,Ref} = u_{Ref2} + \Delta u^*_{LCi} \end{cases} \tag{9.1}$$

As shown in Figure 9.2, the control signals for the ith SM in the upper, middle, and lower switching arms can be obtained by comparing the ith carrier signal u_{Ci} (where $i = 1, 2, \ldots, N$) with the reference signals $u_{Ui,Ref}$, $u_{Mi,Ref1}$, $u_{Mi,Ref2}$, and $u_{Li,Ref}$. For example, the gate signal for the ith upper SM $u_{gUi} = 1$ when $u_{Ui, Ref} > u_{Ci}$, however, the gate signal for the lower sub-module $u_{gLi} = 1$ when $u_{Li, Ref} < u_{Ci}$, and the gate signal for the middle SM $u_{gMi} = 1$ when either of the XOR gate inputs is equal to one.

To verify the validity of the capacitor voltage control scheme presented above, some simulation waveforms of a single-input dual-output half-bridge single-phase DC–AC converter when N = 4 are given in Figure 9.3, which illustrates the first output voltage u_1, the first output current i_1, the second output voltage u_2, the second output current i_2, the capacitor voltages in the upper switching arm, $u_{UC1} \sim u_{UC4}$, the capacitor voltages in the middle switching arm, $u_{MC1} \sim u_{MC4}$, and the capacitor voltages in the lower switching arm, $u_{LC1} \sim u_{LC4}$, respectively. It is found that all of the capacitor voltages can be controlled successfully to the reference value (250 V).

For the single-input dual-output full-bridge single-phase converter and the single-input dual-output three-phase DC–AC converter, the capacitor voltages should not only be balanced in one arm, but also be averaged between different phases. Thus, the closed-loop control signal for both the single-input dual-output single-phase and three-phase DC–AC converters Δu^*_{XjCi} should consider the voltage balancing signal Δu_{XjCi} and the voltage averaging signal Δu_{XCavj} must consider the functions of voltage balancing and voltage averaging, where arm X = U, M, L, phase j = a, b for single-phase DC–AC converter, and j = u, v, w for three-phase DC–AC converter, respectively.

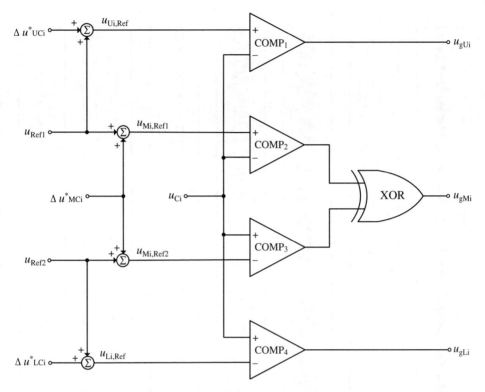

Figure 9.2 Complete control schematic for the single-input dual-output DC–AC converter.

As shown in Figure 9.4, the voltage balancing signal Δu_{XjCi} can be obtained by a similar scheme to that presented in Figure 9.1. In addition, the average voltage of capacitors in one arm u_{XCavj} is subtracted from the reference capacitor voltage U_{CRef}, and their difference value is sent to a PI controller to generate the DC loop current command i_{ZRefj}. The upper arm current i_{Uj} and the lower arm current i_{Lj} are sampled and averaged to obtain the DC loop current i_{Zj}. Then, the difference value between i_{ZRefj} and i_{Zj} is sent to another PI controller to generate the voltage averaging signal Δu_{XCavj}. Finally, the voltage control signal Δu^{*}_{XjCi} for the ith capacitor in arm X of phase j is obtained by dividing the sum of Δu_{XjCi} and Δu_{XCavj} by the capacitor voltage u_{XjCi}.

Similar to Eq. (9.1), the reference signals for the single-input dual-output single-phase and three-phase DC–AC converters when considering the capacitor voltage control are defined by the following equations:

$$\begin{cases} u_{Ui,Refj} = u_{Refj1} + \Delta u^{*}_{UjCi} \\ u_{Mi,Refj1} = u_{Refj1} + \Delta u^{*}_{MjCi} \\ u_{Mi,Refj2} = u_{Refj2} + \Delta u^{*}_{MjCi} \\ u_{Li,Refj} = u_{Refj2} + \Delta u^{*}_{LjCi} \end{cases} \tag{9.2}$$

Figure 9.5 illustrates how to generate the control signals for the ith SM in the upper, middle, and lower switching arms of phase j.

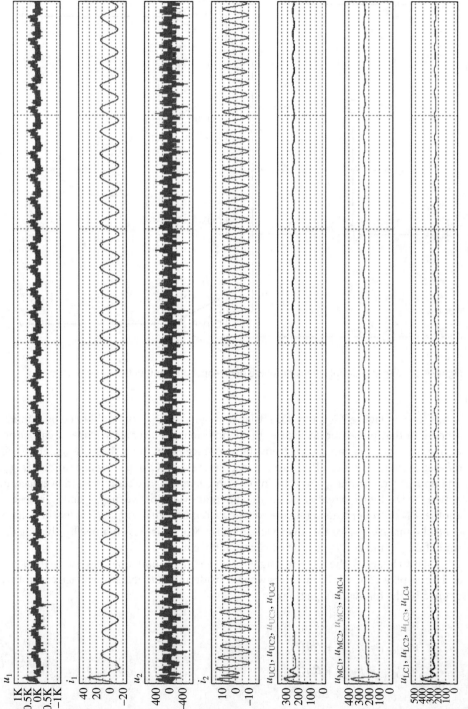

Figure 9.3 Simulation waveforms of the proposed capacitor voltage control scheme for single-input dual-output half-bridge single-phase DC–AC converter.

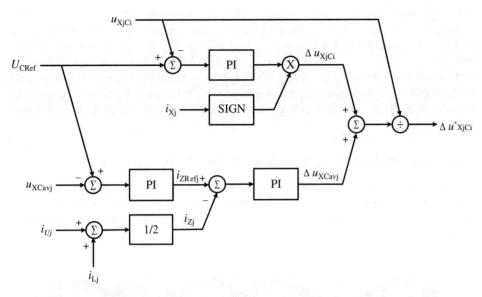

Figure 9.4 Schematic of generating the closed-loop control signal for the single-input dual-output single-phase and three-phase DC–AC converters.

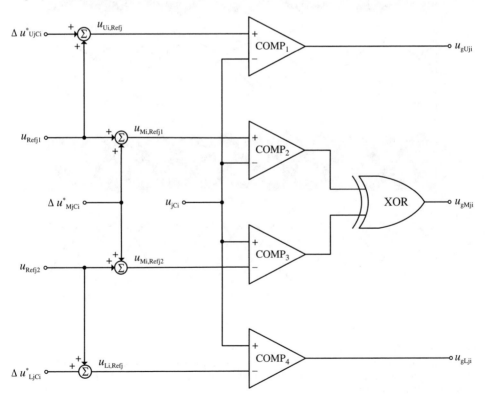

Figure 9.5 Complete control schematic of the single-input dual-output single-phase and three-phase DC–AC converters.

A single-input dual-output three-phase DC–AC converter with N = 8 is used as an example to verify the validity of the proposed control scheme. Figure 9.6a provides the simulation waveforms of the first output line voltage u_{uv1}, u_{vw1}, u_{wu1}, the first output line current i_{uv1}, i_{vw1}, i_{wu1}, the second output line voltage u_{uv2}, u_{vw2}, u_{wu2}, and the second output line current i_{uv2}, i_{vw2}, i_{wu2}. The capacitor voltages in the upper, middle, and lower switching arms of phase u, v, w are illustrated in Figure 9.6b–d, respectively. It is found that all of the capacitor voltages can be controlled successfully to the reference value (250 V).

9.2.2 Single-Phase Multiple-Input Single-Output AC–DC Converter

To balance the capacitor voltages in each arm of the single-phase 2M-arm multiple-input single-output AC–DC converter shown in Figure 4.12, a new control scheme in which each switching arm has an independent modulation signal is presented first [4].

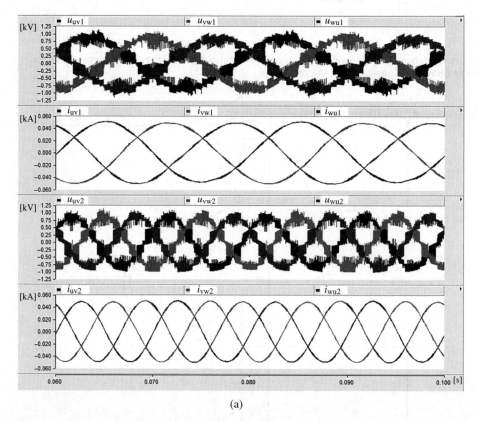

(a)

Figure 9.6 Simulation waveforms of the proposed capacitor voltage control scheme for the single-input dual-output three-phase DC–AC converter. (a) The line voltages and line currents of the loads #1 and #2. (b) SM capacitor voltages of phase u. (c) SM capacitor voltages of phase v. (d) SM capacitor voltages of phase w.

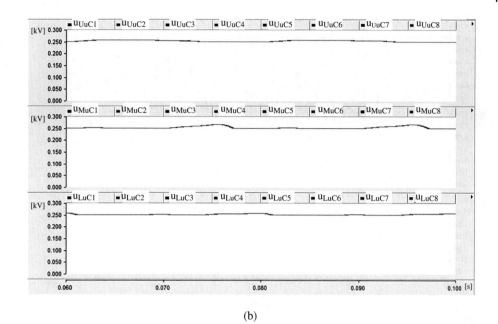

(b)

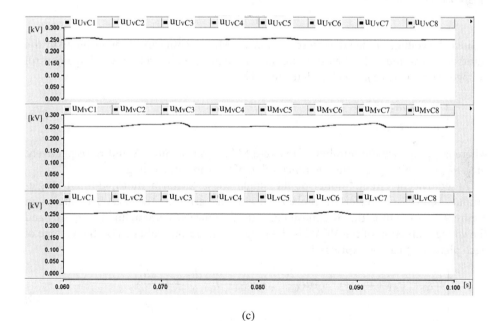

(c)

Figure 9.6 (*Continued*)

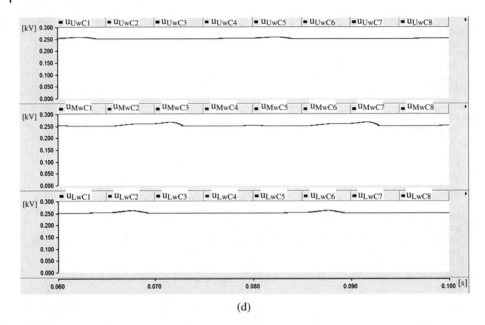

(d)

Figure 9.6 (*Continued*)

Since the voltage of the kth (where $k = 1, 2, \ldots, M$) switching arm is determined by the number of inserted SMs, the kth switching arm can be replaced by the voltage-control voltage source (VCVS), which is determined by

$$\begin{cases} u_{Ak} = m_{Ak} u_{CRef} \\ u_{Bk} = m_{Bk} u_{CRef} \end{cases} \tag{9.3}$$

where m_{Ak}, m_{Bk} are the number of inserted SMs in phase units A and B, respectively, m_{Ak}, $m_{Bk} \in [0, N]$. u_{CRef} is the reference value of the capacitor voltage.

The equivalent circuit model of the single-phase 2M-arm multiple-input single-output AC–DC converter by replacing switching arms to the VCVS is shown in Figure 9.7a. Assuming that the mesh currents in Figure 9.7a are defined by $i_1 \sim i_M$ and the voltage direction of the VCVS is the same as the output voltage U_O, the voltage of each phase unit can be expressed by

$$\begin{cases} U_{AO} = u_{A1} + u_{A2} + \cdots + u_{A(M-1)} + u_{AM} - L_{pa}\dfrac{di_1}{dt} - L_{sa}\dfrac{di_M}{dt} \\ U_{BO} = u_{B1} + u_{B2} + \cdots + u_{B(M-1)} + u_{BM} + L_{pb}\dfrac{di_1}{dt} + L_{sb}\dfrac{di_M}{dt} \end{cases} \tag{9.4}$$

Based on the symmetry of phase units A and B, the jth AC source u_j and its series inductor L_j (where $j = 1, 2, \ldots, M-1$) can be divided into two parts in order to decouple the influence between two phase units, and the corresponding decoupled model is shown in Figure 9.7b.

The voltage of the reference ground GND_j after decoupling is given by

$$u_{GNDj} = \frac{M-j}{M} U_O \tag{9.5}$$

Thus, the switching arm voltages of each phase unit can be obtained by

$$
\begin{cases}
u_{A1} = \dfrac{1}{M}U_O - \dfrac{1}{2}\left[u_1 - L_1\dfrac{d(i_1 - i_2)}{dt}\right] + L_{pa}\dfrac{di_1}{dt} \\[2ex]
u_{A2} = \dfrac{1}{M}U_O + \dfrac{1}{2}\left[u_1 - L_1\dfrac{d(i_1 - i_2)}{dt}\right] - \dfrac{1}{2}\left[u_2 - L_2\dfrac{d(i_2 - i_3)}{dt}\right] \\[2ex]
u_{A3} = \dfrac{1}{M}U_O + \dfrac{1}{2}\left[u_2 - L_2\dfrac{d(i_2 - i_3)}{dt}\right] - \dfrac{1}{2}\left[u_3 - L_3\dfrac{d(i_3 - i_4)}{dt}\right] \\[2ex]
\cdots \\[2ex]
u_{A(M-1)} = \dfrac{1}{M}U_O + \dfrac{1}{2}\left[u_{M-2} - L_{M-2}\dfrac{d(i_{M-2} - i_{M-1})}{dt}\right] \\[2ex]
\qquad\qquad - \dfrac{1}{2}\left[u_{M-1} - L_{M-1}\dfrac{d(i_{M-1} - i_M)}{dt}\right] \\[2ex]
u_{AM} = \dfrac{1}{M}U_O + \dfrac{1}{2}\left[u_{M-1} - L_{M-1}\dfrac{d(i_{M-1} - i_M)}{dt}\right] + L_{sa}\dfrac{di_M}{dt}
\end{cases}
\tag{9.6}
$$

Figure 9.7 Equivalent models of single-phase 2M-arm multiple-input single-output AC–DC converter. (a) Equivalent circuit model. (b) Decoupled circuit model.

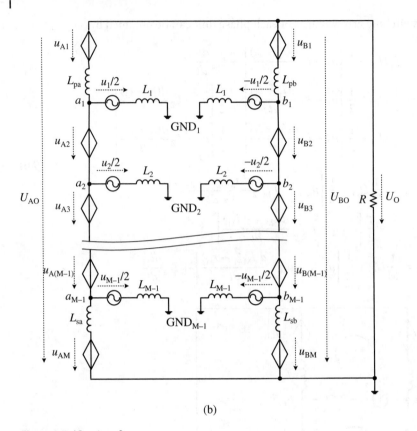

(b)

Figure 9.7 (*Continued*)

$$
\begin{cases}
u_{B1} = \dfrac{1}{M}U_O + \dfrac{1}{2}\left[u_1 - L_1\dfrac{d(i_1 - i_2)}{dt}\right] - L_{pb}\dfrac{di_1}{dt} \\[2mm]
u_{B2} = \dfrac{1}{M}U_O - \dfrac{1}{2}\left[u_1 - L_1\dfrac{d(i_1 - i_2)}{dt}\right] + \dfrac{1}{2}\left[u_2 - L_2\dfrac{d(i_2 - i_3)}{dt}\right] \\[2mm]
u_{B3} = \dfrac{1}{M}U_O - \dfrac{1}{2}\left[u_2 - L_2\dfrac{d(i_2 - i_3)}{dt}\right] + \dfrac{1}{2}\left[u_3 - L_3\dfrac{d(i_3 - i_4)}{dt}\right] \\[2mm]
\cdots \\[2mm]
u_{B(M-1)} = \dfrac{1}{M}U_O - \dfrac{1}{2}\left[u_{M-2} - L_{M-2}\dfrac{d(i_{M-2} - i_{M-1})}{dt}\right] \\[2mm]
\qquad\qquad + \dfrac{1}{2}\left[u_{M-1} - L_{M-1}\dfrac{d(i_{M-1} - i_M)}{dt}\right] \\[2mm]
u_{BM} = \dfrac{1}{M}U_O - \dfrac{1}{2}\left[u_{M-1} - L_{M-1}\dfrac{d(i_{M-1} - i_M)}{dt}\right] - L_{sb}\dfrac{di_M}{dt}
\end{cases}
\tag{9.7}
$$

Based on Eqs. (9.6) and (9.7), the fundamental component of the switching arm voltages should be

$$
\begin{cases}
u_{\text{A1f}} = \dfrac{1}{M}U_{\text{O}} - \dfrac{1}{2}u_1 \\[2mm]
u_{\text{A2f}} = \dfrac{1}{M}U_{\text{O}} - \dfrac{1}{2}(u_2 - u_1) \\[2mm]
u_{\text{A3f}} = \dfrac{1}{M}U_{\text{O}} - \dfrac{1}{2}(u_3 - u_2) \\[2mm]
\cdots \\[2mm]
u_{\text{A(M-1)f}} = \dfrac{1}{M}U_{\text{O}} - \dfrac{1}{2}(u_{M-1} - u_{M-2}) \\[2mm]
u_{\text{AMf}} = \dfrac{1}{M}U_{\text{O}} + \dfrac{1}{2}u_{M-1}
\end{cases}
\tag{9.8}
$$

$$
\begin{cases}
u_{\text{B1f}} = \dfrac{1}{M}U_{\text{O}} + \dfrac{1}{2}u_1 \\[2mm]
u_{\text{B2f}} = \dfrac{1}{M}U_{\text{O}} + \dfrac{1}{2}(u_2 - u_1) \\[2mm]
u_{\text{B3f}} = \dfrac{1}{M}U_{\text{O}} + \dfrac{1}{2}(u_3 - u_2) \\[2mm]
\cdots \\[2mm]
u_{\text{B(M-1)f}} = \dfrac{1}{M}U_{\text{O}} + \dfrac{1}{2}(u_{M-1} - u_{M-2}) \\[2mm]
u_{\text{BMf}} = \dfrac{1}{M}U_{\text{O}} - \dfrac{1}{2}u_{M-1}
\end{cases}
\tag{9.9}
$$

Assuming that the reference signal corresponding to the jth AC input source, $u_{\text{Ref}j}$, is expressed by

$$
u_{\text{Ref}j} = \frac{Mu_j}{U_{\text{O}}}
\tag{9.10}
$$

the reference signals to regulate the switching arm voltages are defined by

$$
\begin{cases}
u^*_{\text{A1,Ref}} = 1 - \dfrac{1}{2}u_{\text{Ref1}} \\[2mm]
u^*_{\text{A2,Ref}} = 1 - \dfrac{1}{2}(u_{\text{Ref2}} - u_{\text{Ref1}}) \\[2mm]
u^*_{\text{A3,Ref}} = 1 - \dfrac{1}{2}(u_{\text{Ref3}} - u_{\text{Ref2}}) \\[2mm]
\cdots \\[2mm]
u^*_{\text{A(M-1),Ref}} = 1 - \dfrac{1}{2}(u_{\text{Ref(M-1)}} - u_{\text{Ref(M-2)}}) \\[2mm]
u^*_{\text{AM,Ref}} = 1 + \dfrac{1}{2}u_{\text{RefM}}
\end{cases}
\tag{9.11}
$$

$$\begin{cases} u^*{}_{B1,Ref} = \dfrac{1}{2}u_{Ref1} \\[2mm] u^*{}_{B2,Ref} = \dfrac{1}{2}(u_{Ref2} - u_{Ref1}) \\[2mm] u^*{}_{B3,Ref} = \dfrac{1}{2}(u_{Ref3} - u_{Ref2}) \\[2mm] \cdots \\[2mm] u^*{}_{B(M-1),Ref} = \dfrac{1}{2}(u_{Ref(M-1)} - u_{Ref(M-2)}) \\[2mm] u^*{}_{BM,Ref} = -\dfrac{1}{2}u_{RefM} \end{cases} \qquad (9.12)$$

As each reference signal is independent after decoupling, the switching arm voltage can be controlled individually. Similar to Section 9.2.1, the capacitor voltage could be balanced by adding an additional closed-loop control signal $\Delta u^*{}_{XkCi}$ to the reference signal of the switching arm. In order to generate $\Delta u^*{}_{XkCi}$, the relationship between the capacitor voltage and the capacitor current (or the switching arm current) should be discussed.

The current distribution of the decoupled single-phase 2M-arm multiple-input single-output AC–DC converter is shown in Figure 9.8, in which i_{uj} represents the jth

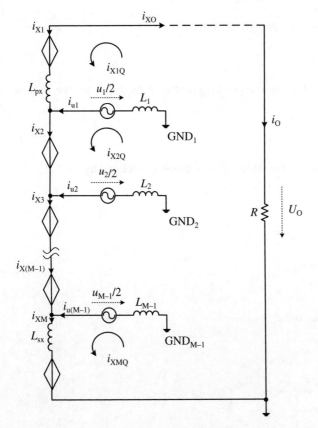

Figure 9.8 Current distribution of the single-phase 2M-arm multiple-input single-output AC–DC converter.

input current, i_{XO} is the output current of phase unit X (X = A, B), and i_{XkQ} is the reactive current of the kth switching arm. According to the energy conservation law, the following equation can be established:

$$i_O = \frac{i_{u1}u_1 + i_{u2}u_2 + i_{u3}u_3 + \cdots + i_{u(M-1)}u_{(M-1)}}{u_O} = i_{AO} + i_{BO} \tag{9.13}$$

However, the reactive capacity is changing all the time during the actual operating process, and results in the reactive current i_{XkQ}. The voltage drop Δu_{Xk} caused by i_{XkQ} can be considered as the voltage drop on the inductors in the current loop, which is expressed by

$$
\begin{cases}
\Delta u_{X1} = L_{px} \dfrac{d(i_{X1Q} - i_{XO})}{dt} + \dfrac{L_1}{2} \dfrac{d(i_{X1Q} - i_{X2Q})}{dt} \\[2mm]
\Delta u_{X2} = \dfrac{L_1}{2} \dfrac{d(i_{X2Q} - i_{X1Q})}{dt} + \dfrac{L_2}{2} \dfrac{d(i_{X2Q} - i_{X3Q})}{dt} \\[2mm]
\Delta u_{X3} = \dfrac{L_2}{2} \dfrac{d(i_{X3Q} - i_{X2Q})}{dt} + \dfrac{L_3}{2} \dfrac{d(i_{X3Q} - i_{X4Q})}{dt} \\[2mm]
\cdots \\[2mm]
\Delta u_{X(M-1)} = \dfrac{L_{(M-2)}}{2} \dfrac{d(i_{X(M-1)Q} - i_{X(M-2)Q})}{dt} + \dfrac{L_{(M-1)}}{2} \dfrac{d(i_{(M-1)Q} - i_{XMQ})}{dt} \\[2mm]
\Delta u_{XM} = L_{sx} \dfrac{d(i_{XMQ} - i_{XO})}{dt} + \dfrac{L_{(M-1)}}{2} \dfrac{d(i_{XMQ} - i_{X(M-1)Q})}{dt}
\end{cases} \tag{9.14}
$$

Obviously, the relationship among Δu_{Xk}, the reference voltage $u_{Xk,Ref}$ of the kth switching arm, and the switching arm voltage u_{Xk} can be determined by

$$u_{Xk,Ref} - u_{Xk} = \Delta u_{Xk} \tag{9.15}$$

Based on Eqs. (9.3), (9.14), and (9.15), the difference between the reference capacitor voltage u_{CRef} and the average capacitor value u_{XkCav} of the kth switching arm is decided by

$$
\begin{cases}
u_{CRef} - u_{X1Cav} = \dfrac{L_{px}}{m_{X1}} \dfrac{d(i_{X1Q} - i_{XO})}{dt} + \dfrac{L_1}{2m_{X1}} \dfrac{d(i_{X1Q} - i_{X2Q})}{dt} = \dfrac{L_{px}}{m_{X1}} \dfrac{di_{X1}}{dt} - \dfrac{L_1}{2m_{X1}} \dfrac{di_{u1}}{dt} \\[2mm]
u_{CRef} - u_{X2Cav} = \dfrac{L_1}{2m_{X2}} \dfrac{d(i_{X2Q} - i_{X1Q})}{dt} + \dfrac{L_2}{2m_{X2}} \dfrac{d(i_{X2Q} - i_{X3Q})}{dt} = \dfrac{L_1}{2m_{X2}} \dfrac{di_{u1}}{dt} - \dfrac{L_2}{2m_{X2}} \dfrac{di_{u2}}{dt} \\[2mm]
u_{CRef} - u_{X3Cav} = \dfrac{L_2}{2m_{X3}} \dfrac{d(i_{X3Q} - i_{X2Q})}{dt} + \dfrac{L_3}{2m_{X3}} \dfrac{d(i_{X3Q} - i_{X4Q})}{dt} = \dfrac{L_2}{2m_{X3}} \dfrac{di_{u2}}{dt} - \dfrac{L_3}{2m_{X3}} \dfrac{di_{u3}}{dt} \\[2mm]
\cdots \\[2mm]
u_{CRef} - u_{X(M-1)Cav} = \dfrac{L_{M-2}}{2m_{X(M-1)}} \dfrac{d(i_{X(M-1)Q} - i_{X(M-2)Q})}{dt} + \dfrac{L_{M-1}}{2m_{X(M-1)}} \dfrac{d(i_{X(M-1)Q} - i_{XMQ})}{dt} \\[2mm]
\qquad\qquad\quad = \dfrac{L_{M-2}}{2m_{X(M-1)}} \dfrac{di_{u(M-2)}}{dt} - \dfrac{L_{M-1}}{2m_{X(M-1)}} \dfrac{di_{u(M-1)}}{dt} \\[2mm]
u_{CRef} - u_{XMCav} = \dfrac{L_{sx}}{m_{XM}} \dfrac{d(i_{XMQ} - i_{XO})}{dt} + \dfrac{L_{M-1}}{2m_{XM}} \dfrac{d(i_{XMQ} - i_{X(M-1)Q})}{dt} \\[2mm]
\qquad\qquad\quad = \dfrac{L_{sx}}{m_{XM}} \dfrac{di_{XM}}{dt} + \dfrac{L_{M-1}}{2m_{XM}} \dfrac{di_{u(M-1)}}{dt}
\end{cases} \tag{9.16}
$$

By integrating Eq. (9.16), we have

$$\int (u_{CRef} - u_{XkCav}) = \frac{L_{Xk} i'_{Xk}}{m_{Xk}}$$

(9.17)

where the fixed switching arm current is expressed by

$$\begin{cases} i'_{X1} = \dfrac{1}{L_{X1}} \left(L_{px} i_{X1} - \dfrac{L_1}{2} i_{u1} \right) \\[2mm] i'_{X2} = \dfrac{1}{L_{X2}} \left(\dfrac{L_1}{2} i_{u1} - \dfrac{L_2}{2} i_{u2} \right) \\[2mm] i'_{X3} = \dfrac{1}{L_{X3}} \left(\dfrac{L_2}{2} i_{u2} - \dfrac{L_3}{2} i_{u3} \right) \\[2mm] \cdots \\[2mm] i'_{X(M-1)} = \dfrac{1}{L_{X(M-1)}} \left(\dfrac{L_{(M-2)}}{2} i_{u(M-2)} - \dfrac{L_{M-1}}{2} i_{u(M-1)} \right) \\[2mm] i'_{XM} = \dfrac{1}{L_{XM}} \left(L_{sx} i_{XM} + \dfrac{L_{(M-1)}}{2} i_{u(M-1)} \right) \end{cases}$$

(9.18)

Therefore, by sensing all input AC currents $i_{u1} \sim i_{u(M-1)}$, the first and last switching arm current i_{X1} and i_{XM}, the voltage averaging signal Δu_{XkCav} can be obtained based on Eqs. (9.17) and (9.18), while the voltage balancing signal Δu_{XkCi} can be obtained by the same method presented in Section 9.2.1. In summary, the capacitor voltage control signal Δu^*_{XkCi} for the ith capacitor in the kth switching arm of phase unit X can be generated by the schematic shown in Figure 9.9 [5].

Combining the capacitor voltage control signal Δu^*_{XkCi} with the switching arm reference signal $u^*_{Xk,Ref}$ determined by Eqs. (9.11) and (9.12), the complete reference signal for the ith SM in the kth switching arm of phase unit X is obtain by

$$u_{XkCi,Ref} = u^*_{Xk,Ref} + \Delta u^*_{XkCi}$$

(9.19)

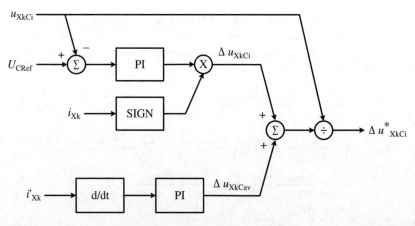

Figure 9.9 Schematic of generating the closed-loop capacitor control signal for the single-phase 2M-arm multiple-input single-output AC–DC converter.

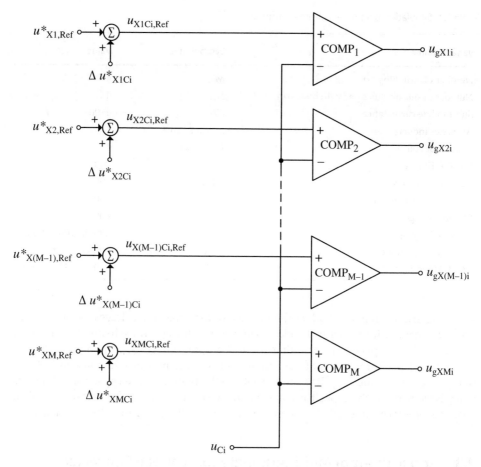

Figure 9.10 Complete control schematic of the single-phase 2M-arm multiple-input single-output AC–DC converter.

Similar to the SM control method presented in Section 3.2.2, the control signal, u_{gXki}, for the ith SM in the kth switching arm of phase unit X, can be obtained by comparing the ith carrier signal u_{Ci} to the reference signal $u_{\text{XkCi,Ref}}$, as illustrated in Figure 9.10.

To verify the validity of the decoupling control scheme and the capacitor voltage control scheme presented above, a single-phase 2M-arm multiple-input single-output AC–DC converter prototype is built in MATLAB Simulink®, while the simulation parameters are given in Table 9.1 [4].

When only the decoupling control scheme is applied, some simulation waveforms are given in Figure 9.11a, including three input voltages $u_1 \sim u_3$, the switching arm voltages of phase unit A $u_{\text{A1}} \sim u_{\text{A4}}$, and the DC output voltage U_O. Though three input voltages have different magnitude, phase shift or frequency, it is found that the output voltage U_O can be controlled successfully to the required value ($U_O = 800$ V).

To illustrate the function of the proposed closed-loop capacitor voltage control scheme, the load R changes from 400Ω to 200Ω at $t = 3$ s, which will change the

Table 9.1 Simulation parameters of the prototype.

Variable name	Symbol/unit	Value
Number of switching arms	M	4
Number of sub-modules per switching arm	N	4
Sub-module capacitance	$C/\mu F$	4700
AC series inductance	$L_1, L_2, L_3/mH$	1
Phase inductance	$L_{px}, L_{sx}/mH$	1
Load resistance	R/Ω	400
Carrier frequency	f/Hz	1000
Reference capacitor voltage	u_{CRef}/V	100
Output voltage	U_O/V	800
Input AC voltage #1	u_1/V	$220\sin(100\pi t + 2\pi/3)$
Input AC voltage #2	u_2/V	$200\sin(100\pi t)$
Input AC voltage #3	u_3/V	$200\sin(120\pi t - 2\pi/3)$

switching arm currents and affect the balance of the capacitor voltages. Some simulation waveforms without/with the closed-loop capacitor control scheme are shown in Figure 9.11b,c, respectively, including the DC output voltage U_O, the output current i_O, the first capacitor voltages of the first and fourth switching arm in phase unit A and B, u_{A11}, u_{B11}, u_{A41}, and u_{B41}. It is found that the fluctuation of the capacitor voltages is really large without the capacitor voltage control scheme, but can be better controlled near the reference value (100 V) when the capacitor voltage control scheme is used.

9.3 Applications of Multi-terminal High-voltage Converter

Based on the introduction in preceding chapters, the proposed multi-terminal high-voltage converters can perform single-phase and three-phase AC–DC or DC–AC conversion successfully through appropriate control. Thus, the multi-terminal converter has potential application prospects in the field of renewable energy power generation and industrial motor drive. In this section, some application frameworks based on multi-terminal converters will be introduced, in which multiple wind turbines can deliver energy to the AC or DC bus, and multiple industrial motors can be driven by a single AC or DC power source.

9.3.1 Multiple Wind Turbines and DC Bus

Based on the three-phase 3M-arm multiple-input single-output AC–DC converter shown in Figure 4.15, the framework which incorporates multiple wind turbines into the DC bus is illustrated in Figure 9.12, where all AC terminals of the converter are connected to M − 1 wind turbines and the DC terminal is considered as the DC bus. Through appropriate control, the electric power generated by wind turbines can be delivered to the DC bus by only one multi-terminal high-voltage converter.

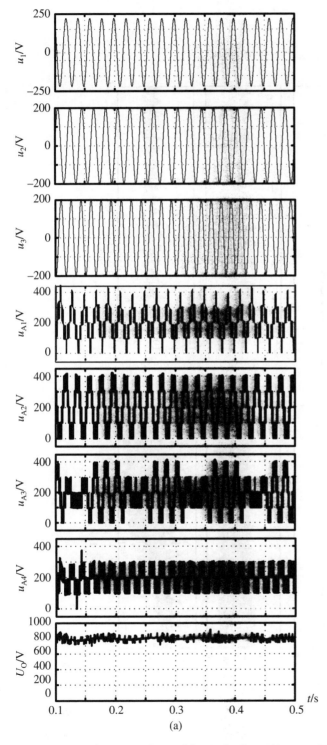

Figure 9.11 Simulation waveforms of the single-phase 2M-arm multiple-input single-output AC–DC converter. (a) With the decoupling control scheme only. (b) Without the capacitor voltage control scheme. (c) With the capacitor voltage control scheme.

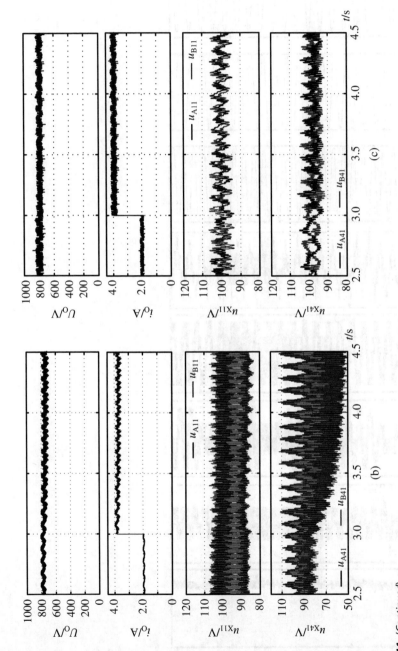

Figure 9.11 (Continued)

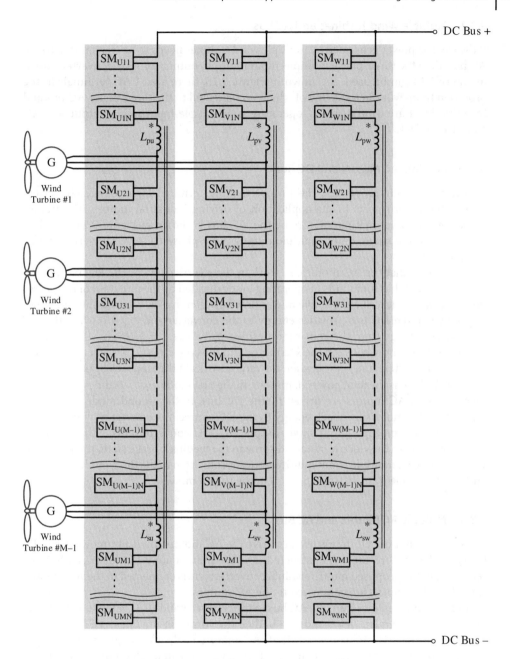

Figure 9.12 Framework of connecting multiple wind turbines to a DC bus.

9.3.2 Multiple Wind Turbines and AC Bus

If the electric power generated by multiple wind turbines needs to be transmitted to the AC bus, then the three-phase multiple-input multiple-output AC–AC converter shown in Figure 5.13 can be used. As shown in Figure 9.13, there are M AC terminals in the proposed framework in total, one of which is connected to the three-phase AC bus and the other M−1 are wind turbines, operating as a multiple-input single-output AC–AC converter in fact.

9.3.3 Multiple AC Motors and DC Bus

According to their properties, the proposed multi-terminal high-voltage converters have potential prospects in the application of driving industrial motors. The framework used to drive multiple motors by the DC bus is shown in Figure 9.14, which is derived from the single-input multiple-output three-phase DC–AC converter shown in Figure 3.17.

Considering that the AC motor can work in the electric state or the feedback brake state, the system has four kinds of operating mode decided by the direction of the power flow. Use p_{DC} to describe the active power flow between the DC bus and the converter, $p_{DC} > 0$ when the DC bus provides energy to the converter and $p_{DC} < 0$ when the DC bus absorbs energy from the converter. Besides, p_k (where k = 1, 2, ..., M) represents the active power flow between the kth motor and the converter, $p_k > 0$ when the motor works in the electric state, $p_k < 0$ when the motor works in the feedback brake state, and $p_{Motor} = \sum_{k=1}^{M} p_k$ is the total power of motors. In the first mode, $p_{DC} > 0$ and $p_k > 0$, which means that all AC motors are driven by the DC bus. In the second mode, $p_k < 0$ and $p_{DC} < 0$, all AC motors feedback energy to the DC source; the converter in Figure 9.14 changes to be a multiple-input single-output AC–DC converter at this time. If some motors operate in the electric state and some in the feedback brake state, then the converter becomes a multiple-input multiple-output converter. In the third mode, $p_{Motor} > 0$ and $p_{DC} > 0$, while $p_{Motor} < 0$ and $p_{DC} < 0$ in the fourth mode.

9.3.4 Multiple AC Motors and AC Bus

If the AC bus not the DC bus is used to drive the AC motors, then a framework similar to that in Figure 9.13 can be constructed. As shown in Figure 9.15, there are M+1 AC terminals in the proposed multi-terminal converter; one is connected to the AC bus and the other M terminals are connected to the AC motors.

When the power flows from the AC bus to the converter and all AC motors work in the electrical state, that is $p_{AC} > 0$ and $p_k > 0$, then the converter operates as a single-input multiple-output AC–AC converter. If all AC motors feedback energy to the AC bus, then $p_k < 0$ and $p_{AC} < 0$, and the converter turns out to be a multiple-input single-output AC–AC converter. If some AC motors work in the electric state and some in the feedback brake state, then the converter becomes a multiple-input multiple-output AC–AC converter. The direction of power flow in the AC bus depends on p_{Motor}: if $p_{Motor} > 0$ then $p_{AC} > 0$, while if $p_{Motor} < 0$ then $p_{AC} < 0$.

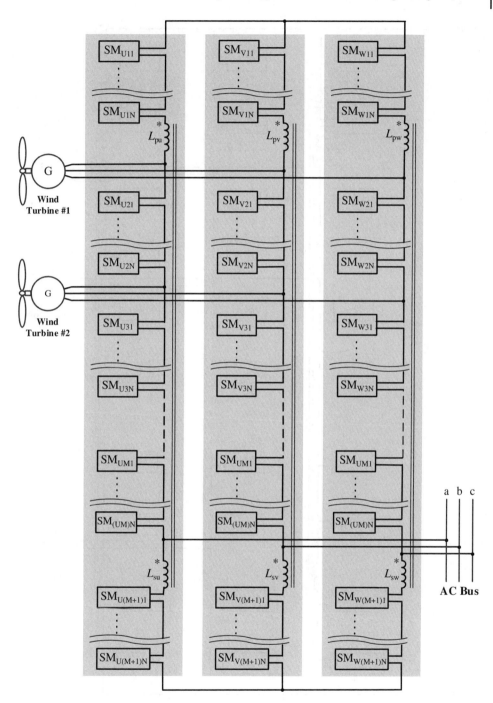

Figure 9.13 Framework of connecting multiple wind turbines to an AC bus.

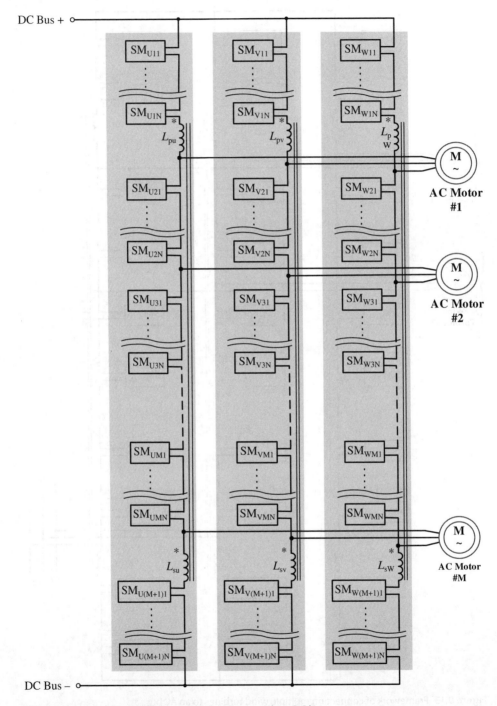

Figure 9.14 Framework of driving multiple AC motors by a DC bus.

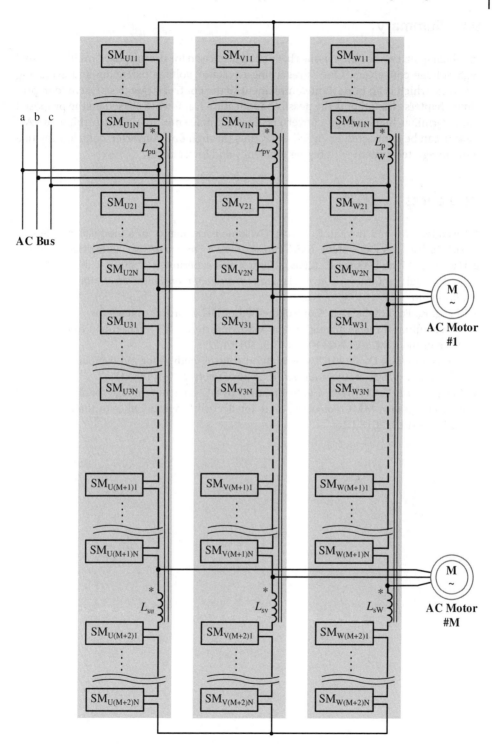

Figure 9.15 Framework of driving multiple AC motors by an AC bus.

9.4 Summary

In this chapter, two common issues have been discussed for the proposed multi-terminal high-voltage converters. One is regarding capacitor voltage balancing and averaging schemes, which is an important complement to the control strategies presented in previous chapters. The other is the possible application frameworks based on the proposed multi-terminal high-voltage converters. For example, multiple wind turbines or AC motors can be connected to an AC or DC bus through one converter, contributing to a more integrated system with higher efficiency and lower cost.

References

1 Saeedifard, M. and Iravani, R. (2010). Dynamic performance of a modular multilevel back-to-back HVDC system. *IEEE Transaction on Power Delivery* 25 (4): 2903–2912.
2 Hagiwara, M. and Akagi, H. (2009). Control and experiment of pulse width modulated modular multilevel converters. *IEEE Transaction on Power Electronics* 24 (7): 1737–1746.
3 Zhang, B., Fu, J., Qiu, D. Y. Control method of DC capacitor voltage for double-output three switching-groups MMC inverter. State Intellectual Property Office of the P.R.C., ZL201410145647.3, 2017.4.12.
4 Qin, L. and Qiu, D. Y. (2017). Research on control method for multi-input single-phase MMC rectifier. *Journal of Power Supply* 15 (3): 168–175.
5 Zhang, B., Fu, J., Qiu, D. Y. Control method of DC capacitor voltage for six- and nine switching-groups MMC converter. State Intellectual Property Office of the P.R.C., ZL 20141039319.2, 2018.1.16.

Index

Multi-terminal High-voltage Converter, First Edition. Bo Zhang and Dongyuan Qiu.
© 2019 John Wiley & Sons Singapore Pte. Ltd. Published 2019 by John Wiley & Sons Singapore Pte. Ltd.